T0305926

"Some years ago a student of sustainability put forward the notion that as far as the future goes, either we shall have a sustainable society or we shall have no society at all. Building on the concept of 'Dual Process,' Ashley Colby provides a specific case study, giving us a roadmap of how to replace the dysfunctional institutions of capitalism with institutions that are sane and sustainable—what she calls 'shadow structures.' The importance of such specific information is clear: while most Americans are intensely embroiled in meaningless and vapid discussions of Democrats vs. Republicans, they are oblivious to the real drama going on, namely the ongoing demise of capitalism and the coming of a post-capitalist society. One can only hope that Dr. Colby's work will serve to wake them up."

Morris Berman, independent scholar, Mexico

"Subsistence food production is one of the most dynamic and socially progressive elements of the contemporary food movement. Colby puts the subsistence food production of South Chicago under the lens in this theoretically sophisticated ethnography of resistance at the margins of consumer capitalism."

Richard Wilk, Distinguished Professor Emeritus,
Department of Anthropology, Indiana University

"Ashley Colby's *Subsistence Agriculture in the US* is a work for our time: the late capitalist era of COVID-19, climate change, economic disruption, and precarious labor. Focusing on the growth of subsistence food production in the United States, her book sees the emergence of such 'shadow structures' as the rational response of growing numbers of people to a period of increasing ecological and social malaise. The result is a highly original approach to the environmental sociology of long-term, transformative social change."

John Bellamy Foster, Professor of Sociology,
University of Oregon; author of *The Return of Nature:
Socialism and Ecology* (2020)

"We at New Dream are committed to the work of imagining a new kind of future that transforms the systems we have in place to one that improves well-being for all people and the planet. Dr. Ashley Colby's book gives us an example of one such transformation taking place, right in plain sight, if we can only shift our thinking to see it as important. By talking to those practicing subsistence agriculture in and around the South Side of Chicago, Dr. Colby reveals to us that it may be the rural hunters and fishermen, the urban backyard gardeners and canners, and the suburban households keeping chickens and goats who may be at the forefront of developing a new American Dream. Dr. Colby shows us how it is not only the act of production, but the connections people make across race, class, gender and geography, that may be laying this important groundwork for a different kind of economic and political structure."

Guinevere Higgins,
Director of Strategic Partnerships, New Dream

Subsistence Agriculture in the US

Focusing on ethnography and interviews with subsistence food producers, this book explores the resilience, innovation and creativity taking place in the subsistence agricultural industry in America.

To date, researchers interested in alternative food networks have often overlooked the somewhat hidden, unorganized population of household food producers. *Subsistence Agriculture in the US* fills this gap in the existing literature by examining the lived experiences of people taking part in subsistence food production. Over the course of the book, Colby draws on accounts from a broad and diverse network of people who are hunting, fishing, gardening, keeping livestock and gathering and looks in depth at the way in which these practical actions have transformed their relationship to labor and land. She also explores the broader implications of this pro-environmental activity for social change and sustainable futures.

With a combination of rigorous academic investigation and engagement with pressing social issues, this book will be of great interest to scholars of sustainable consumption, environmental sociology and social movements.

Ashley Colby earned her PhD focusing on environmental sociology from Washington State University in 2018. Ashley's dissertation research is on subsistence food production as a potentially revolutionary act that is in some ways attempting to develop a post-capitalist future. Ashley got her MA in sociology at WSU in 2013, and her BA in Cinema and Media Studies at the University of Chicago in 2007. She has travelled to over 30 countries on five continents. Ashley is currently interested in and passionate about the myriad creative ways in which people are forming new social worlds in resistance to the failures of late capitalism and resultant climate disasters. As a qualitative researcher she tends to focus on the informal spaces of innovation.

She is currently pursuing research projects based in Uruguay, where she has recently founded Rizoma Field School for experiential learning on the area of sustainability and agroecology.

Routledge-SCORAI studies in sustainable consumption

This series aims to advance conceptual and empirical contributions to this new and important field of study. For more information about The Sustainable Consumption Research and Action Initiative (SCORAI) and its activities please visit www.scorai.org.

Series Editors:
Maurie J. Cohen, New Jersey Institute of Technology, USA
Halina Szejnwald Brown, Professor Emerita of Clark University, USA
Philip J. Vergragt, Professor Emeritus of Delft University, Netherlands

Titles in this series include:

Social Innovation and Sustainable Consumption
Research and action for societal transformation
Edited by Julia Backhaus, Audley Genus, Sylvia Lorek, Edina Vadovics, Julia M. Wittmayer

Power and Politics in Sustainable Consumption Research and Practice
Edited by Cindy Isenhour, Mari Martiskainen and Lucie Middlemiss

Local Consumption and Global Environmental Impacts
Accounting, trade-offs and sustainability
Kuishuang Feng, Klaus Hubacek and Yang Yu

Subsistence Agriculture in the US
Reconnecting to work, nature and community
Ashley Colby

For more information about this series, please visit: https://www.routledge.com/Routledge-SCORAI-Studies-in-Sustainable-Consumption/book-series/RSSC

Subsistence Agriculture in the US

Reconnecting to Work, Nature and Community

Ashley Colby

Routledge
Taylor & Francis Group

LONDON AND NEW YORK

earthscan
from Routledge

First published 2021
by Routledge
2 Park Square, Milton Park, Abingdon, Oxon OX14 4RN

and by Routledge
52 Vanderbilt Avenue, New York, NY 10017

Routledge is an imprint of the Taylor & Francis Group, an informa business

British Library Cataloguing-in-Publication Data
A catalogue record for this book is available from the British Library

Library of Congress Cataloging-in-Publication Data
A catalog record has been requested for this book

ISBN: 978-0-367-45872-0 (hbk)
ISBN: 978-1-003-02588-7 (ebk)

Typeset in Times New Roman
by MPS Limited, Dehradun

To Patrick, Isabel, Vivian and Lucia

Contents

List of tables   x
Series editors' introduction   xi
Preface   xv

1   Introduction: Building shadow structures at the
    crisis of industrial capitalism   1

2   Subsistence agriculture in South Chicago   20

3   Guiding theories: Social problems, emergent
    solutions   37

4   Who are subsistence food producers in Chicago?
    Meanings across class of alienation and viscerality   52

5   "It connects me to the Earth:" Marginalized
    environmentalism and a resistance to
    capitalist logic   67

6   "Without the garden we never would have met
    him:" Practitioner networks as post-capitalist
    shadow structures   86

7   Conclusion: "We've got to find a solution"   112

    Index   122

Tables

2.1 Descriptive statistics of sample   31
3.1 Weber's characteristics of pre-modern
     and modern eras   42

Series editors' introduction

Philip Vergragt, Lucie Middlemiss, Halina Brown

> *We learn from our gardens to deal with the most urgent question of the time: How much is enough?*
>
> Wendell Berry (1977, 56)

As we write, Corona virus is creating havoc in how we conduct our daily lives – restricting our mobility, causing fear of contagion through physical contact and food contamination, and resulting in losses of household income. Ashley Colby's book about urban subsistence agriculture in the US is particularly timely. Readers might be surprised to learn that subsistence food production (SFP) is widely practiced in urban Chicago, especially on its infamous South Side, which is plagued by multiple problems of poverty, violence, drug use, and lack of access to services that more affluent neighborhoods take for granted.

As we find out, small-scale food production has existed in urban and suburban areas of Chicago throughout its history, including agriculture, hunting, fishing, and bartering all of which are culturally acceptable ways to address food insecurity. While in the past SFP was largely driven by economic necessity, today the motivation is often the distrust of the dominant agri-food system and what that does to food quality, as well as underlying feelings of alienation from the prevalent system of production and consumption. But there is a paradox here: people who are savvy enough to recognize the systemic problems in the public sphere choose to address them through personalized and private solutions.

Colby's book provides us with a rich empirical story of SFP in Chicago. It draws on 60 semi-structured interviews with subsistence food producers and approximately 120 hours of participant observation, mainly at the site of food production for each household, as well as at public events, such as fairs, markets, expos, and so on. She tells the stories of subsistence food producers in the Chicago area, how they think, feel, act, and interact, what they learn and how they connect. The book offers important insights on class: Colby's subjects (whom she labels as either "upper class" or "lower class") engage in a range of ecological practices, and the author explores their motivations, using the concept of "ecological embeddedness" as well as considering the larger ramifications of SFP for social change.

Not surprisingly, Colby finds that food producers in her sample demonstrate a deep care for the environment. Their proximity to, and stakes in the ecological systems that produce their food lead them to practice composting, water use reduction, promoting biodiversity through diverse agroecology, organic pest/disease management, and, in the case of hunters and fishermen, conservation-minded land management. These practices contrast starkly with industrial food production practices that they do not incorporate into their small-scale production. Notably, while many lower class SFP-ers reject the ideology of what they refer to as "political environmentalism of the higher middle classes" they in fact behave in a more ecologically friendly manner than many of their upper-class counterparts.

This study documents the blurring of class boundaries among subsistence food producers, the way in which food production gives space for conversations across these boundaries. Individual producers form connections to each other through sharing and exchange of information and providing mutual support. The author describes a mid-winter fair in South Chicago where an upper-class white woman and a working-class black woman engage in an easy and lively conversation about practical SFP problems, something that rarely happens otherwise because their paths do not typically cross. And when a local politician wants to pass a prohibition against backyard chicken coops, a successful political action mobilizes food producers as a single community, irrespective of class and race.

Another noteworthy finding is that the interactions within the SFP community sometimes lead to the emergence of groups committed to social change. Through cooperation and mutual support, the individualized response to the systemic problem of industrialized food production – to become a subsistence farmer – evolves into a collective endeavor.

Colby's work provides a significant contribution to a burgeoning critical literature on community and environment (reviewed in Taylor Aiken et al. 2017), detailing relatively constructive class relations in Chicago, in contrast to the more combative class and race relations observed elsewhere (Anantharaman et al. 2019; Huddart Kennedy et al. 2019; Grossmann and Creamer 2016). As such this is a welcome follow-up to the previous book in this Routledge-SCORAI Studies in Sustainable Consumption, *Power and Politics in Sustainable Consumption Research and Practice* (Isenhour et al. 2019), which features a political approach to sustainable consumption. Indeed this book makes two central points that are echoed here: first, that we must pay attention to the politics of identity and difference; and second, that there is a need to explore how the political economy influences mundane acts of consumption.

Colby also develops a fruitful political economy approach through the concept of shadow structures: social organizations that adopt an alternative logic to that of late capitalism to meet perceived needs, and exist parallel to (not in direct conflict with) the dominant system (Berman 2017). The shadow economy concept offers a Marxist perspective on social change and remind us of the "interstitial processes" introduced by Eric Olin Wright (Wright 2010). It also introduces a more political perspective on the 'niches' of the transitions literature (Geels and Schot 2007) as well as providing insights for scholars working on prefigurative politics, and a helpful framework for those engaged in thinking about just transitions in social and environmental change. As such, there is a natural link here to another earlier book in Routledge-SCORAI Studies in Sustainable Consumption, *The Coming of Post-Consumer Society: Theoretical Advances and Policy Implications,* which examines the cases of ongoing social innovations toward sustainable consumption through the lenses of four widely accepted theories of social change (Cohen et al. 2017).

Can we speak of an incipient social movement? The author makes an argument in the affirmative, but the readers should make their own judgment. If it is, one of its notable characteristics is that, contrary to the general understanding of social movements, SFP transcends class boundaries.

Works cited

Anantharaman, M. et al. 2019. "Who Participates in Community-based Sustainable Consumption Projects and Why Does It Matter? A Constructively Critical Approach." *Power and Politics in Sustainable Consumption Research and Practice*. Eds. C. Isenhour, M. Martiskainen and L. Middlemiss. Abingdon: Routledge: 178.

Berman, Morris. 2017. "Dual Process: The Only Game in Town." In *Are We There Yet?* Brattleboro, VT: Echo Point Books. Essay #27. http://morrisberman.blogspot.com/2016/05/dual-process-only-game-in-town.html

Cohen, M., H.S. Brown and P.J. Vergragt. 2017. *Social Change and the Coming of Post-Consumer Society: Theoretical Advances and Policy Implications*. Routledge https://www.routledge.com/Social-Change-and-the-Coming-of-Post-consumer-Society-Theoretical-Advances/Cohen-Brown-Vergragt/p/book/9781138642058

Geels, F. and Schot, J. (2007) "Typology of Sociotechnical Transition Pathways." *Research Policy* 36(3):399–417.

Grossmann, M. and E. Creamer. 2016. "Assessing Diversity and Inclusivity within the Transition Movement: An Urban Case Study." *Environmental Politics* 26(1):161–182.

Huddart Kennedy, E. et al. 2019. "Eating for Taste and Eating for Change: Ethical Consumption as a High-Status Practice." 98(1):381–402.

Isenhour, C., et al. (2019). *Power and Politics in Sustainable Consumption Research and Practice*. Abingdon: Routledge.

Taylor Aiken, G. et al. (2017). "Researching Climate Change and Community in Neoliberal Contexts: An Emerging Critical Approach." *Wiley Interdisciplinary Reviews: Climate Change* 8(4):e463.

Wright, E. (2010) *Envisioning Real Utopias*. New York: Verso.

Preface

I am writing this preface during the ongoing COVID-19 pandemic. The uncertainty and chaos across the world is palpable. As the outbreak has laid bare some of the most fundamental failures of the global capitalist system to provide basic care and necessities to our societies, the findings of this book come into starker relief.

Back in 2015/6, when I conducted the interviews for this book with subsistence food producers in Chicago, many of them reported to me that they were driven to self-produce food as a result of the perceived risk and uncertainty at the margins of crisis. They suggested that it was this feeling of risk, of a fundamentally unstable social system, that made them want to take part in a tangible, practical solution that could have immediate results in providing the basic security of food.

It is unclear, in this moment, the prospect for global food supply chains and how exactly the restrictions of social distancing will lead to challenges in accessing basic goods. We know from early reporting that community gardens are ramping up food producing operations to hedge against potential closures of grocery stores amid the pandemic (Wharton 2020), and that farmers are destroying milk, eggs and vegetables as restaurants abide mandatory shutdowns (Newman and Bunge 2020). It is in moments such as these, when the failures of the social system get laid so unambiguously bare, that individuals start to rethink, reimagine and create.

Certainly, the political response in the United States does not inspire confidence, and it is unfortunate that individuals must turn to private solutions rather than rely on their government to provide in ways that is beyond the scope of individual solutions (something I explore in chapter six). Yet, the U.S. government is using this crisis to advance disaster capitalism (Klein 2007) that strengthens monopolies and puts more money in the hands of the wealthy and powerful to the detriment of individual thriving. Some argue the grift of the corporate elite in the legislative bailout response to COVID-19 is set to be much worse than what happened in the 'golden parachute' era of the 2008 recession (Taibbi 2020).

It is important to see COVID-19 not as an aberration, but as part of a steady stream of failures that have been playing out for decades, impacting less fortunate populations first and saving the relatively insulated privileged populations for later in the process of its demise. There are major events that indicate these failures of the system to thrive: the 9/11 terrorist attacks, Hurricane Katrina, the 2008/9 recession. However, there are also myriad small failures: the creep of food deserts, the lack of social safety net, or the uptick in deaths of despair among the working class (Case and Deaton 2020). Indeed, the subsistence food producers in my sample were keen to these many failures years ago. It is only with this more widespread sense of crisis, where more members of society are impacted, that these discussions about the crises inherent in the system come to the fore.

Some, like historian Morris Berman whose Dual Process Theory I use throughout the book, claim that these failures were inherent in American values of hustling and wealth creation above all else (2014). Others, drawing on Marxist theory, argue that these crises are built into the architecture of the capitalist system, which destroys itself as it squeezes wealth from the working class (Harvey 2011). In either case, I make the argument throughout the book that we are already in the process of developing the system that will succeed when the current system fails completely, and we must pay attention to those at the margins who are exploring these alternatives, like subsistence food producers.

I am reminded of a saying, often referred to as a curse: *may you live in interesting times.* Indeed, living through the failure of capitalism and the emergent next system is an interesting time to live. I can imagine it is like living through the end of the Roman Empire, as Christianity began to take a foothold as the next world order. We are at an *interesting* moment, and it is up to us to make the world we want to see come next.

My hope is that readers of this book can follow the example of the community of subsistence food producers for lessons in creating the new world we hope to see. I implore researchers, politicians and activists to see the importance of these communities who are actively creating new worlds at the margins and in the midst of crisis, and to amplify the work they are doing, and to join them.

Works Cited

Berman, Morris. 2014. *Why America Failed: The roots of imperial decline.* Createspace Independent Publishing Platform.

Case, Anne and Angus Deaton. 2020. *Deaths of Despair and the Future of Capitalism.* Princeton, NJ: Princeton University Press.

Harvey, David. 2011. *The Enigma of Capital: and the Crises of Capitalism.* London: Oxford University Press.

Klein, Naomi. 2007. *The Shock Doctrine: The Rise of Disaster Capitalism.* Toronto: Random House of Canada.

Newman, Jesse and Jacob Bunge. 2020. "Farmers dump milk, break eggs as Coronavirus restaurant closings destroy demand." *The Wall Street Journal.* Accessed 9 April 2020.

Taibbi, Matt. 2020. "The SEC Rule and Destroyed the Universe: How the Coronavirus is creating a political opportunity to overturn one of the worst practices of the kleptocracy era." <https://taibbi.substack.com/p/the-sec-rule-that-destroyed-the-universe> Accessed 13 April 2020.

Wharton, Rachel. 2020. "'If all the stores close, we need food:' Community gardens adapt to the pandemic." *The New York Times.* Accessed 10 April 2020.

1 Introduction
Building shadow structures at the crisis of industrial capitalism

> The most systematic and comprehensive organic and living alternative to existing hegemonies comes not from the ivory towers or the factories but from the fields.
>
> – Rajeev Patel (2007, 90)

It was a below freezing Chicago February day. In the middle of the afternoon the sun was hidden behind thick winter cloud cover. The snow was blowing across the parking lot of a Chicago Public High School on the city's far South Side, where I parked and made my way quickly inside. As I walked through the front door, I was greeted by a pen of milking goats inside a mobile enclosure with bowls of water, dry food and a bale of hay. As I walked toward the goats I found that the smell was, surprisingly, not too strong: earthy and sweet. They were extremely curious, and I felt as though any minute one of them would make the leap out of this weak enclosure and start wandering into some obscure corner of the school. As I looked down the main hallway, I saw a number of tables set up with rabbits, quail, guinea fowl, chickens and roosters, and even ducks who were inside a plastic gate with a children's pool.

This was the fourth annual Urban Agriculture Livestock Expo hosted by Chicago's Advocates for Urban Agriculture, a coalition whose stated goal is to "support and expand sustainable agriculture in the Chicago area, from home- and community-based growing to market gardens and small farms" (2018, 1). The expo was a free annual event meant to allow those who keep animals for food production to share information with fellow animal food producers as well as anyone interested enough to come to the event. As I looked around, I saw people of all ages, races, and genders mingling to chat about keeping animals in and around Chicago for food production – both for meat and animal products.

A middle-aged white man walked up to a young black woman at the table in front of the ducks:

MAN: So, why ducks instead of chickens?
WOMAN: Well, for one thing they're smarter. And the eggs taste better. But you do have to figure out the water.
M: Water?
W: Yes, ducks need some kind of pond to be happy and healthy.
M: Oh, and how do you do that exactly in this Chicago winter?
W: Well, first you need to figure out your space...

(Field notes. February 14, 2015)

The conversation continued.

I felt as if I had come across a hidden population that is taking part in a private activity of subsistence food production (SFP[1]). Individuals at this event represented diversity across dimensions of social difference such as age, race and gender.[2] Despite this seemingly vast diversity, people at the Expo connected over their specific interest in common: subsistence food production practices, in this case with a special focus on animal agriculture. In this seemingly mundane space, inside a high school in Chicago in the middle of winter, I saw something that is both potentially transformative yet full of contradictions.

Sociologists have been interested in social, economic and political responses to modernity since the founding of the discipline. Scholars such as Marx and Weber (1930) studied the ways in which industrialization changed the structure of modern life. Recent scholarship has focused on the ways in which late capitalism is laden with contradictions and looming crises (Harvey 2017; Roberts 2009) – economic, political, and environmental. Academics from a variety of backgrounds have taken notice of these crises (Beling et al. 2017; Fischer-Kowalski and Haberl 2007), and argue that our global society may be on the precipice of another major transformation analogous to the industrial revolution, whether we actively embrace it or not (Polanyi 1944; Leonard 2011). This new era of crises is also acknowledged by "risk society" theorist Ulrich Beck, who makes the claim that our society is in a state of metamorphosis, which "implies a much more radical transformation in which the old certainties of modern society are falling away and something quite new is emerging" (2016, 1).

Economic sociologist Karl Polanyi's (1944) seminal theory of the double movement argues that as market forces work to commodify all aspects of life – including land/nature and labor/human beings – counter movements arise that seek to de-commodify these resources. It is a process by which "society defends itself against domination by the self-regulating market" (Evans 2008, 271). Much of this theory is in the lineage of Marx and Weber – especially in the discussion of alienation and disenchantment of the industrial era and the resultant push to counter alienation and enact re-enchantment (Berman 1981; Foster 1999; Foster and Holleman 2012).

Taking as an assumption that the global capitalist system under which most of the world operates is now facing deeply structural changes, some scholars are exploring the alternatives that are simultaneously arising as a result of these crises (Berman 2017; McClintock 2014; Wright 2010). I define these as *shadow structures*, which can be any sort of individual or group social organization that adopts an alternative logic to that of late capitalism, meets presented needs, and exists parallel to (not in direct conflict with) the hegemonic industrial system. Some examples of shadow structures include: alternative currencies, tool or seed libraries, alternative modes of transportation, downshifters, voluntary simplicity, non-monetary exchanges/bartering, gift economies, freecycling or dumpster diving, among others.

This story is unlike the one that is the primary focus of sustainable consumption literature. We hear from social scientists that environmental social problems are seemingly insurmountable (Brulle and Pellow 2006; Catton 1994; Davidson and Andrews 2013; Deitz and Rosa 1994; Frickel and Freudenberg 1996; Molotch 1976; Pellow 2000; York, Rosa and Deitz 2003). Scholars of sustainable consumption tend to look for solutions within the community of predominantly white, educated, wealthy environmentalists of the Global North (Boli and Thomas 1997; Boucher 2017; Frank, Hironaka and Schofer 2000; Franzen and Mayer 2010; Inglehart 1995). I move beyond the myopic solutions that advocate slight changes in individual consumer behavior, or those that hope that technological efficiencies will lead to widespread transformation (for more on this critique, see Akenji 2014; Geels et al. 2015; Isenhour, Martiskainen and Middlemiss 2019; Lorek and Fuchs 2013; Middlemiss 2018; ORourke and Lollo 2015). When we are faced with the failure of an economic system that is fundamentally set up to exploit land and people, we must look to individuals and communities exploring transformational solutions.

This book seeks to provide next steps for readers at an impasse from the bombardment of difficult news and seemingly insurmountable crises by demonstrating the innovative, tangible and meaningful ways individuals are already combating the failures of late capitalism and simultaneously creating a new future. Further, the diversity of the people whose lives I explore in this book help to outline a sustainable future that is inclusive, varied, and innovative from people who are rarely invited into the conversation on solutions to environmental problems. However, not every aspect of this shadow structure development represents a perfect solution, so I also explore the paradoxes, contradictions and challenges that arise in this messy, complicated process of making new social structures.

By drawing on the concept of shadow structures from Berman's Dual Process Theory, I am situating this book in a school of thought that includes Marx, Polanyi, Braudel and the Annales School, Wallerstein, and others whose main focus is the dialectical forces of history and economy. Instead of making prescriptions about what *should* happen (for example, how we should consume better), I am instead describing what I have observed in my data (that is, what people are already *doing*) through the lens of the current historical moment, which includes the messy end of capitalism and the emergent alternatives to it.

In some ways, the meanings of the subsistence producers I observe are particular to their own personal life courses, including the specificities of their time and place. In other ways, they are representative of this moment in history, and the meanings and processes of those living in it (Burawoy 1998; Small 2009). Despite the overlapping social crises unfolding (Beck 2016; Beling et al. 2017; Fischer-Kowalski and Haberl 2007), there are surprising ways in which many members of this diverse SFP community are showing signs of resilience, defined as the ability to adapt to major changes (King 2008).

This research is an in-depth exploration of the meanings and processes brought to the act of subsistence food production in urban, suburban and rural field sites within the larger Chicago metropolitan area. I went into the field wanting to know more about these people who decided to produce a significant amount of food for their own consumption. What drove them to take part in this activity, sometimes seen as laborious and difficult? I was curious to know how their self-reported behaviors and thoughts interacted with academic conceptions of environmental identities and actions. I was also driven to understand the complex interplay between the seemingly private nature of SFP and the potential impact on the more political or public realm.

Marginalized environmentalism

One lens through which to think about mitigating negative environmental effects is sustainable consumption, a term that has been used across academic disciplines and within policy circles for the past couple of decades. Sustainable consumption has often been co-opted by mainstream policy makers to focus on ecological modernization, which includes: increasing efficiencies or renewable technologies, putting power in the hands of individual consumer-citizens without considering limitations of power or structure on the ability to act in the marketplace, or market-based solutions that advocate slight changes in pricing or incentives (Isenhour et al. 2019). Each of these perspectives have frustrated researchers and activists of sustainable consumption, who have instead suggested that it is our fundamental social structures that are the cause of the problem, and that "durable solutions will require social pressure and cooperation, if not social transformation" (Isenhour et al. 2019, 5).

Several scholars of sustainable consumption are already exploring versions of shadow structures from the sharing economy to small-scale experiments to socio-technical niches. Schor and Wengronowitz (2017) find that some aspects of the sharing economy represent structures that mirror business-as-usual capitalism, especially among for profit sectors. However, among those participating in the sharing economy – in both the for profit and non-profit sectors – they have also found an emergent eco-habitus (Carfagna et al. 2014) that values localism, building community, less corporate structures and more sustainability. Although it is unclear the extent to which this culture of eco-habitus is expanding, the authors argue that a more radical, progressive vision is being explored by early adopters in the sharing economy movement.

Sahakian (2017) situates her study of small-scale experiments directly in a Polanyian double movement framework by looking at ways in which the sharing economy can go beyond self-interest and profit maximization to include characteristics such as social cohesion and environmental sustainability in what she calls the solidaristic sharing economy. Like Schor and Wengronowitz (2017), Sahakian sees both characteristics of the capitalist market economy and the solidaristic sharing economy playing out within these sharing initiatives. She argues that we must continue to pay attention to how institutions can foment versions of a solidaristic sharing economy that promote community collaboration, equality of exchange, and are not co-opted by more market capitalistic structures.

Katz-Gerro, Cvetičanin, and Leguina (2017) look at how sustainable lifestyles can arise specifically in response to failures within the economic system more broadly. Exploring individuals and households in Southeastern Europe who are facing a mismatch between their social and economic resources and their (perceived) needs,

Katz-Gerro et al. (2017) find that those most impacted by financial crisis adopt changes in behavior that limit mainstream consumerism and promote sustainability. Some of the adopted practices include: postponing purchasing of non-necessary goods, recalibrating perceived consumer needs, pooling consumption through sharing or bartering, developing the ability to self-provision, and generally moving toward individual or household production when marketplace consumption became inaccessible. The authors provide an excellent example here of how shadow structures develop out of the failures of the economic system to provide. Individuals then turn to the dual response of adapting their expectations of consumer behavior, and "making do" in practical, actionable ways through self-production or sharing.

Rob Hopkins is the architect of the concept of Transition Towns, which takes as an assumption that oil prices will make business-as-usual economic structures impossible, and imagines how to build communities not reliant upon cheap, unlimited fossil fuels (Hopkins 2014). Belgian economist Bernard Lietaer promotes the idea of alternative currencies and time banks (Lietaer 2001; Lietaer, Arnsperger and Goerner 2012). There are hundreds of examples of such currencies across the United States, the oldest of which is called Ithaca Hours, with notes valuing $1/10^{th}$ of an hour through two hours, with the one hour note valuing close to $10 U.S. dollars. Similarly to the findings of Katz-Gerro et al. (2017) discussed earlier in the chapter, Berman suggests that it is in places most affected by the failures of neoliberal capitalism, like in Spain where austerity measures were rolled out as a response to the economic crisis, where explorations of alternative currencies are most thriving (2017).

Jeff Ferrell explores the culture of dumpster diving as a critique of our consumer capitalist culture wherein material throughput leads to salvageable goods being sent to landfills (2008). In the case of each of these phenomena, the literature has begun to explore alternatives emerging through mainly descriptive analysis. Much of this work is fundamentally lacking in a robust theoretical framework that helps us to understand *why* these small-scale experiments might be arising at this point in history, how they relate to the larger social structure, and what is their significance in terms of impacting larger society.

Voluntary simplicity and downshifting have been explored by scholars of sustainable consumption as potentially transformative. Each provide promising possibilities for new ways of relating to work, free time, community and consumption (Dobson 2003; Jackson 2005; Kennedy, Krahn and Krogman 2013). However, many of these studies and solutions tend to focus on a specific kind of community in the Global North: one that is white, highly educated, left-leaning, and urban (Franzen and Mayer 2010) and therefore represents a certain set of norms, structures, and degrees of social freedom (Middlemiss 2018). In fact, in a recent piece Schor and Wengronowitz (2017) acknowledge the widespread understanding (Stack 1974) that working-class, poor communities and communities of color take part in alternative sharing economies of reciprocity and interdependence, but for their study insist on focusing on what they call the "new sharing economy" which includes individuals with high cultural capital who share among strangers using new digital technologies. In order to begin to explore the tremendous task of imagining an entirely different world economic and social system, we must first acknowledge that dimensions of social difference (e.g. race, class, gender, geography) and social structures impact the ability to enact certain kinds of transformational solutions (see Chapter 4). Beyond that, however, I argue that we as researchers should be inclusive and sample to maximize a diverse range of communities so we can understand certain shadow structure phenomena beyond relatively well-studied "environmentalists."

The way in which I explore emergent alternatives is through inquiry into the specific act of subsistence food production (SFP), which I have defined as producing at least fifty percent of one's food needs in the high season of production. Because I think it is important to understand the phenomenon of SFP in a holistic way, I include diversity in my sample of class, race, gender, geography and age. The comprehensive study of individual food production for self-consumption has produced little data. Major nationwide studies within the United States include trade association surveys through which we know that *at least fifty percent* of American households are taking part in some form of food gardening (National Gardening Association 2009; Garden Writer's Association 2013).

Social science disciplines such as geography, anthropology and urban planning have done more comprehensive work in the study of individual or small-scale food production. However, much of this work has a bias toward highly visible farms and community gardens as well as a focus on urban populations. For example, in the discipline of geography, there has been a surge in interest in the study of urban food production. Researchers have looked at urban farms (Pulighe and Lupia 2016), or farming in places like rooftops (Saha and Eckelman 2017) and vacant lots (Hara et al. 2013). Geographers have also looked at socially conscious acts such as guerilla gardening (Hardman and Larkham 2014), planned edible green infrastructure (Russo et al. 2017), or foraging in urban environments (McLain et al. 2014). Researchers have looked at the challenges of urban livestock (Blecha and Leitner 2014), including beekeeping (Moore and Kosut 2013).

One population that has gained much attention both by academic researchers and non-governmental organizations is community gardeners (for a review see Guitart, Pickering and Byrne 2012). Community gardening has been shown to have a number of positive effects including promoting community cohesion (Guitart et al. 2012); producing healthier soils, when compared to industrial agriculture (Edmondson et al. 2014); as well as providing substantial amounts of quality food for gardeners and others that receive excess produce, "dwarf[ing] urban farmers in the volume and value of food they produce" (Vitiello and Wolf-Powers 2014, 10–11), which challenges the notion that individual gardeners do not produce enough food to significantly impact their diet. Much of this research has been exploratory and descriptive, giving us a sense of the landscape of practices, but often lacking in theory.

Within the social sciences, researchers look at the effect of class on the outcomes of participation in alternative food networks. One set of academic work has criticized aspects of the alternative food movement as coming from a privileged positionality, claiming that "the movement's predominantly white and middle-class character suggests that it may itself be something of a monoculture" (Alkon and Agyeman 2011, 2). Other authors argue that members of the middle-class food sovereignty movement may be producing food sustainably at the expense of less privileged groups' ability to produce their own food in a sustainable way (Mares and Pena 2011). One example in anthropologist Heather Paxon's work on American artisanal cheesemaking finds that it is indeed privilege that allows certain populations to choose financial sacrifice for work in high food craft (2013), thus gaining cultural capital in the process. Several authors have found that despite other positive social and environmental outcomes, one negative outcome of privileged populations taking part in urban agriculture is the way in which they display their labor as a form of conspicuous consumption, thus reaffirming the logic of late capitalism (Boltanski and Chiapello 2005; Mincyte and Dobernig 2016).

Taking up these critiques, a subset of researchers has been focusing on self-provisioning practices as a means of combating inequality. Studies have looked at the practice of food self-provisioning in a diverse array of low-income settings, both urban and rural (Block, Chavez and Allen 2011; Levkoe 2006; Sherman 2009; Winne 2010) and across diverse populations including recent immigrants (Gottlieb and Joshi 2010; Mares and Pena 2011) and Native Americans (Morales 2011;

Norgaard, Reed and Van Horn 2011; Winne 2010). Among low-income populations, participation in food production has been demonstrated to promote social resilience (Carolan and Hale 2016) by building social capital (Guitart et al. 2012; Teig et al. 2009; Veen 2015), developing interpersonal skills among youth engaged in community gardening (Ober Allen, Alaimo and Elam 2008), and providing wage-earning opportunities (Ferris et al. 2001; Kaufman and Bailkey 2000). Household food production has been linked to positive health outcomes, especially, though not exclusively, for low-income communities (Freeman et al. 2012; Kremer and DeLiberty 2011; Metcalf and Widener 2011; Wakefield et al. 2007) by fostering healthy eating through access to foods that may otherwise be inaccessible (D'Abundo and Carden 2008; McCormack, Laska and Larson 2010).

In this book, I make the argument for inclusivity and diversity in exploring subsistence food production populations, so as to be able to explore the similarities and differences among groups that are often left out of the literature of solutions within sustainable consumption (e.g., low cultural capital, working class, and communities of color). There is a small but burgeoning literature that finds pro-environmental *behaviors* (whether or not these populations hold environmentalist *identities*) among communities ignored or left out of the conversation on pro-environmental futures. This literature includes examples in the United States such as urban communities of color growing food as a form of political resistance (Donald et al. 2010; Jarosz 2000; Maye, Holloway and Kneafsey 2007; White 2011). Less privileged communities are using food self-provisioning to advance specific political claims such as food sovereignty (Block et al. 2011; Gottlieb and Joshi 2010; Morales 2011), which focuses on community or localized control of food production, and overcoming barriers to accessing culturally valued and nutritious foods.

Researchers and major non-governmental organizations are paying attention to the myriad potential positive social outcomes of alternative food networks. The United Nations recently released a report suggesting small-scale agriculture is one of the most impactful and greatest resources to combat the already deleterious effects of climate change, noting that there are already 2.5 billion small-scale agriculturalists around the world (UNEP 2013). This report claims global small-scale agriculture can provide ecosystems services, strengthen local communities (especially those in poverty), sequester carbon and provide food with a much smaller carbon footprint than industrial agriculture. The World Bank also has a report that calls for the development of small-scale sustainable agriculture as one of the most powerful tools for poverty reduction globally (World Bank 2008).

Again, with an urban bias, the discipline of geography has an uptick in interest on how urban agriculture can address problems such as food access, ecological degradation, economic blight, and health issues (McClintock 2010; Clancy and Ruhf 2010). I argue that by moving beyond urban spaces and bringing this substantive area into dialogue with a constellation of Marxist theory there is much more research to be accomplished exploring these emerging alternative food systems and their effect on myriad social and ecological phenomena.

Prominent researchers (Beck 2016; Buttel 2000; Urry 2010) have been calling for more work in the embedded, localized meanings and processes of those exploring *responses* to the myriad environmental social problems documented by researchers. I take up this call by developing an understanding of this marginalized population of subsistence food producers that is theoretically grounded in Marxist theory, situating the findings in this moment as capitalist crises play out and alternatives emerge (Berman 2017; Foster 1999; Foster and Holleman 2012; Polanyi 1944).

Overview of findings

In Chapter 2, I situate the book in Chicago, on the South Side of the city, in the South Suburbs and in the rural area further south. I discuss my methodology, describe my sample, and give some historical context. In the following chapters, I explore the dialectical *process* of taking part in SFP. In each chapter I explore a different aspect of SFP from culture to environment to social and political. In Chapter 3, I explore in much more detail the theory I will be relying on for the remainder of the book. I discuss how, according to Berman (2017), the development of shadow structures plays out, and how this fits into classical Marxist and Weberian understandings of the relation between humans and nature in industrial capitalism. I then rely on recent theoretical advances made in the field of geography to demonstrate how shadow structures are complex and therefore develop *paradoxically* (Galt et al. 2014; McClintock 2014), and argue that we must embrace this complexity as researchers to more fully understand this phenomenon.

In Chapter 4, I explore the significant meanings that, paradoxically, cut across class and other dimensions of difference. These shared meanings center around the idea of alienation from work (Marx 1990; Foster 1999) and the enjoyment of the viscerality, or immediate physicality, in the production of food (Hayes-Conroy 2011). This chapter begins to give us a sense of what is drawing people to self-produce food, and the ascribed meanings they place on the work they do and how it makes them feel connected. The data suggest that the desire to counter the alienation experienced in the era of late capitalism is a frame that cuts across dimensions of social difference and lays the groundwork for the larger social and political implications found in the following chapters of this book.

I build on this finding in Chapter 5 by exploring self-reported identities and behaviors related to environmentalism. Typically, sociologists have focused on people who hold explicitly environmentalist identities as a way to explore the solutions put forward by this subgroup. However, this chapter upends this research strategy by finding that reported pro-environmental behaviors are shared by both SFPers who identify as environmentalists and those who do not. I use the lens of ecological embeddedness (Mincyte and Doebernig 2016; Whiteman and Cooper 2000), or proximity to one's natural environment, as a way to understand how these behaviors develop. I add to the literature by finding the significant ways that SFPers, whether or not they hold environmentalist identities, are making positive choices and connecting to nature through the act of producing food. Just as in Chapter 4 I find the desire to counter alienation through work and viscerality, in this chapter I suggest SFPers have a desire to address the problems of alienation through connection with nature. The implication here is that researchers must move beyond simply studying self-identified environmentalists when exploring emergent pro-environment social solutions.

I then move from the cultural meanings and the environmental practices to the social and political activities of SFPers in Chapter 6. Like the preceding chapters, the results of this chapter can be paradoxical. On the one hand, SFPers are enacting a neoliberal, individualized solution to their perceived problems. Yet, unexpectedly, I find that SFPers are also countering alienation by making connections through communities of practice. These communities help to develop civil society (Putnam 2000) by drawing on the strength of both weak and strong ties (Granovetter 1983). I argue that the results of the development of these communities of practice is the beginning of the formation of shadow structures that closely resemble pre-modern social institutions described by Weber (1930) (see Table 3.1). These shadow structures provide horizontal networks of social action that have been and can be deployed for social change (in this study only on a hyper-local scale), and they also represent alternative economic or social relationships that exist in parallel to hegemonic relations of late capitalism (Harvey 2005).

In this book I would like to draw attention to populations at the margins of research on emergent alternatives to capitalism and explore the way in which these populations ought to be in conversation with mainstream sustainable consumption research and policy. Rather than looking at well-funded NGOs, self-identified environmentalists or directly at high-level federal or international policies looking to combat environmental destruction (like the many failed climate change meetings and non-binding agreements), researchers ought to pay attention to people like Marty,[3] a white, conservative rural Illinois farmer who disdains climate change politics but cares deeply for conservation of wild animals and ecosystems. Or George, a lower class rural black man who is growing most of his own food, making his own biodiesel, and teaching these practices to his low-income community mostly consisting of people of color. These populations marginalized by the search for a sustainable, post-capitalist future may be the ones experimenting and innovating in important ways by both 1) countering alienation through connecting with physical work, nature, and community, and 2) developing alternative economic, political and social arrangements that can provide options for parallel social institutions in places where hegemonic social institutions fail. I will explore the surprising, contradictory and illuminating phenomena of this population of subsistence food producers in Chicago, with important implications for both future research as well as the future of our global society.

Notes

1 Subsistence food production (SFP) has been defined in my sample as a population of people who have self-reported to produce at least 50 percent of their food needs in the high season of production. As the anthropological literature shows (Apfel-Marglin 1997; McMichael 2012), around the world subsistence production is almost never responsible for all individual food needs. Instead, most SFPers globally weave together a complex set of practices both self-producing for consumption and also earning a wage to purchase goods (McMichael 2012).

2 During my ethnographic observations, I was able to take notes of presented race, age and gender of participants. I did speak with a portion of participants at the event (between 20 and 30 percent) and could decipher some information in terms of class and geography. From what I gathered, there was some variability in class, with more middle- and upper-middle class people represented than working class. For geography, participants I spoke with and for whom I overheard geographic details, there was greater representation of urban and suburban populations, but I spoke with at least five people from rural areas, and at least 3 of the presenters who brought animals came from rural areas as well. I noticed ages from small children through elderly adults – with an average age somewhere around

40 years old. The primary race I noticed present was white, but significant portions of the attendees were people of color, with African American being the most highly represented. Gender seemed quite evenly split among presenters, and among attendees there were slightly more males than females present.

3 All names are pseudonyms.

Works cited

Advocates for Urban Agriculture. 2018. "About AUA." (https://auachicago. org/) Retrieved May 30, 2018.

Akenji, L. 2014. Consumer scapegoatism and limits to green consumerism. *Journal of Cleaner Production*, 63: 13–23.

Alkon, Alison Hope and Julian Agyeman. 2011. "Introduction: The Food Movement as a Polyculture." pp. 1–20. In *Cultivating Food Justice: Race, Class and Sustainability*. Eds. Alison Hope Alkon and Julian Agyeman. Cambridge, MA: MIT Press.

Apfel-Marglin, F. 1997. "Counter-development in the Andes." *Ecologist*, 27(6): 221.

Beck, Ulrich. 2016. *The Metamorphosis of the World: How Climate Change is Transforming our Concept of the World*. New York: Polity Press.

Beling, Adrian, Julien Vanhulst, Federico Dermaria and Jerome Pelenc. 2017. "Discursive Synergies for a 'Great Transformation' towards Sustainability: Pragmatic Contributions to a Necessary Dialogue between Human Development, Degrowth and Buen Vivir." *Ecological Economics*. doi: 10.1016/j.ecolecon.2017.08.025.

Berman, Morris. 1981. *The Reenchantment of the World*. Ithaca NY: Cornell University Press.

Berman, Morris. 2017. "Dual Process: The Only Game in Town." in *Are We There Yet?* Brattleboro, VT: Echo Point Books. Essay #27.

Blecha, Jennifer and Helga Leitner. 2014. "Reimagining the Food System, the Economy, and Urban Life: New Urban Chicken-keepers in US Cities." *Urban Geography*, 35(1): 86–108.

Block, Daniel R., Noel Chavez and Erika Allen. 2011. "Food Sovereignty, Urban Food Access, and Food Activism: Contemplating the Connections through Examples from Chicago." *Agriculture and Human Values*, 29(2): 203–215.

Boli, John and George M. Thomas. 1997. "World Culture in the World Polity: A Century of International Non-Governmental Organization." *American Sociological Review*, 62(2): 171–190.

Boltanski, L. and E. Chiapello. 2005. *The New Spirit of Capitalism*. London: Verso.

Boucher, J.L. 2017. "Culture, Carbon and Climate Change: A Class Analysis of Climate Change Belief, Lifestyle Lock-in, and Personal Carbon Footprint." *Socijalna Ekologija*, 25(1): 53–80.

Bourdieu, Pierre. 1984. *Distinction: A Social Critique of the Judgement of Taste*. Cambridge, MA: Harvard University Press.

Brulle, Robert and David Pellow. 2006. "Environmental Justice: Human Health and Environmental Inequalities." *Annual Review of Public Health*, 27: 103–124.

Burawoy, Michael. 1998. "The Extended Case Method." *Sociological Theory*, 16(1): 4–33.

Buttel, Frederick. 2000. "World Society, the Nation-State and Environmental Protection: Comment on Frank, Hironaka and Schofer." *American Sociological Review*, 65(1): 117–121.

Carfagna, Lindsey B., Emilie A. Dubois, Conner Fitzmaurice, Monique Ouimette, Juliet B. Schor, Margaret Willis and Thomas Laidley. 2014. "An Emerging Eco-habitus: The Reconfiguration of High Cultural Capital Practices among Ethical Consumers." *Journal of Consumer Culture*, 14(2): 158–178.

Carolan, Michael and J. Hale. 2016. "'Growing' Communities with Urban Agriculture: Generating Value Above and Below Ground." *Community Development*. doi:10.1080/15575330.2016.1158198.

Catton, William. 1994. "Foundations of Human Ecology." *Sociological Perspectives*, 37(1): 75–95.

Clancy, Kate and Kathryn Ruhf. 2010. "Is Local Enough? Some Arguments for Regional Food Systems." *Choices*, 25(1).

D'Abundo, M. and A. Carden. 2008. "Growing Wellness: The Possibility of Collective Wellness through Community Garden Education Programs." *Community Development*, 39: 83–94.

Davidson, Debra J. and Jeffrey Andrews. 2013. "Not All About Consumption." *Science*, 339: 1286.

Dietz, Thomas and Eugene A. Rosa. 1994. "Rethinking the Environmental Impacts of Population, Affluence and Technology." *Human Ecology Review*, 1: 277–300.

Dobson, A. 2003. *Citizenship and the Environment*. Oxford, UK: Oxford University Press.

Donald, B. et al., 2010. "Re-regionalizing the Food System?" *Cambridge Journal of Regions, Economy and Society*, 3(2): 171–175.

Edmondson, Jill L., Zoe G. Davies, Kevin J. Gaston and Jonathan R. Leake. 2014. "Urban Cultivation in Allotments Maintains Soil Qualities Adversely Affected by Conventional Agriculture." *Journal of Applied Ecology*, 1–10.

Evans, Peter. 2008. "Is an Alternative Globalization Possible?" *Politics and Society*, 36(2): 271–305.

Ferrell, Jeff. 2008. "Happiness is a Warm Dumpster." Core Connections Lecture Series. University of New England. Biddeford, Maine.

Ferris, J., C. Norman and J. Sempik. 2001. "People, Land and Sustainability: Community Gardens and the Social Dimension of Sustainable Development." *Social Policy and Administration*, 35: 559–568.

Fischer-Kowalski, M. and H. Haberl. 2007. *Socioecological Transitions and Social Change: Trajectories of Social Metabolism and Land Use.* In "Advances in Ecological Economics," series editor: Jeroen van den Bergh. Cheltenham, UK and Northampton, USA: Edward Elgar.

Foster, John Bellamy. 1999. "Marx's Theory of Metabolic Rift: Classical Foundations for Environmental Sociology." *The American Journal of Sociology*, 105(2): 366–405.

Foster, John Bellamy and Hannah Holleman. 2012. "Weber and the Environment: Classical Foundations for a Post-exemptionalist Sociology" *American Journal of Sociology*, 117(6): 1625–1673.

Frank, David John, Ann Hironaka and Evan Schofer. 2000. "The Nation-State and the Natural Environment over the Twentieth Century." *American Sociological Review*, 6(1): 96–116.

Franzen, Axel and Reto Mayer. 2010. "Environmental Attitudes in Cross-national Perspective: A Multilevel Analysis of the ISSP 1993 and 2000." *European Sociological Review*, 26(2).

Freeman, C. et al. 2012. "My garden is an expression of me": Exploring Householders' Relationships with their Gardens." *Journal of Environmental Psychology*, 32(2): 135–143.

Frickel, Scott and William Freudenburg. 1996. "Mining the Past: Historical Context and the Changing Implications of Natural Resource Extraction." *Social Problems*, 43(4): 444–466.

Galt, Ryan E., Leslie C. Gray, and Patrick Hurley. 2014. "Subversive and Interstitial Food Spaces: Transforming Selves, Societies, and Society–Environment Relations through Urban Agriculture and Foraging." *Local Environment*, 19(2): 133–146.

Garden Writer's Association. 2013. *Garden Trends Research Report: Edible Gardening Survey.* Retrieved April 20, 2013 from http://kaga.wsulibs.wsu.edu/ebooks/13_winter_survey.pdf.

Geels, F.W., A. McMeekin, J. Mylan, and D. Southerton. 2015. "A Critical Appraisal of Sustainable Consumption and Production Research: The Reformist, Revolutionary and Reconfiguration Positions." *Global Environmental Change*, 34: 1–12.

Gottlieb, R. and A. Joshi. 2010. *Food Justice.* Cambridge, MA: MIT Press.

Granovetter, Mark. 1983. "The Strength of Weak Ties." *Sociological Theory*, 1: 201–233.

Guitart, D., C. Pickering, and J. Byrne. 2012. "Past Results and Future Directions in Urban Community Gardens Research." *Urban Forestry & Urban Greening*, 11: 364–373.

Hara, Y., A. Murakami, K. Tsuchiya, A.M. Palijon and M. Yokohari. 2013. "A Quantitative Assessment of Vegetable Farming on Vacant Lots in an Urban Fringe Area in Metro Manila: Can it Sustain Long-term Local Vegetable Demand?" *Applied Geography*, 41: 195–206. doi:10.1016/j.apgeog.2013.04.003.

Hardman, Michael and Peter Larkham, 2014. *Informal Urban Agriculture: The Secret Lives of Guerilla gardeners.* New York: Springer International Publishing.

16   *Introduction*

Harvey, David. 2005. *A Brief History of Neoliberalism*. New York: Oxford University Press.

Harvey, David. 2017. *Marx, Capital, and the Madness of Economic Reason*. New York: Oxford University Press.

Hayes-Conroy, Jessica, 2011. "School Gardens and 'Actually Existing' Neoliberalism." *Humboldt Journal of Social Relations*, 33(1/2): 64–96.

Hopkins, Rob. 2014. *The Transition Handbook: From Oil Dependency to Local Resilience*. UIT Cambridge Limited.

Inglehart, Ronald. 1995. "Public Support for Environmental Protection: Objective Problems and Subjective Values in 43 Societies." *Political Science and Politics*, 28(1): 57–72.

Isenhour, Cindy, Mari Martiskainen and Lucie Middlemiss. 2019. *Power and Politics in Sustainable Consumption Research and Practice*. Philadelphia: Routledge. [VitalSource Bookshelf]. Retrieved from https://bookshelf.vitalsource.com/#/books/9781351677301/.

Jackson, T. 2005. "Live Better by Consuming Less? Is there a 'Double Dividend' in Sustainable Consumption?" *Journal of Industrial Ecology*, 9: 19–36.

Jarosz, L. 2000. "Understanding Agri-food Networks as Social Relations." *Agriculture and Human Values*, 17(3): 279–283.

Katz-Gerro, Tally, Predrag Cvetičanin and Adrian Leguina. 2017. "Consumption and Social Change: Sustainable Lifestyles in Times of Economic Crisis." In *Social Change and the Coming of Post-Consumer Society*. Philadelphia: Routledge.

Kaufman, J. and M. Bailkey. 2000. *Farming Inside Cities: Entrepreneurial Urban Agriculture in the United States*. Working paper, Lincoln Institute of Land Policy, Cambridge, MA.

Kennedy, E.H., H. Krahn, and N.T. Krogman. 2013. "Downshifting: An Exploration of Motivations, Quality of Life, and Environmental Practices." *Sociological Forum*, 28(4): 764–783.

King, Christine A. 2008. "Community Resilience and Contemporary Agri-ecological Systems: Reconnecting People and Food, and People with People." *Systems Research and Behavioral Science*, 25: 111–124.

Kremer, P. and T.L. DeLiberty. 2011. "Local Food Practices and Growing Potential: Mapping the Case of Philadelphia." *Applied Geography*, 31(4): 1252–1261.

Lamont, Michele. 1992. *Money, Morals, and Manners: The Culture of the French and the American Upper-Middle Class*. Chicago: University of Chicago Press.

Leonard, Annie. 2011. "Global Change: By Disaster or By Design?" Presentation at Tulane University, New Orleans, LA. October 3, 2011.

Levkoe, C.Z. 2006. "Learning Democracy through Food Justice Movements." *Agriculture and Human Values*, 23: 89–98.

Lietaer Bernard, Christian Arnsperger, Sally Goerner and Stefan Brunnhuber. 2012. *Money and Sustainability: The Missing Link*. London: Triarchy Press.

Lietaer, B. 2001. *The Future of Money: A New Way to Create Wealth, Work and a Wiser World*. London: Random House.

Lorek, S. and D. Fuchs. 2013. "Strong Sustainable Consumption Governance–Precondition for a Degrowth Path?" *Journal of Cleaner Production*, 38: 36–43.

Mares, Teresa M. and Devon G. Pena. 2011. "Environmental and Food Justice: Toward Local, Slow and Deep Food Systems." pp. 197–219. In *Cultivating Food Justice: Race, Class and Sustainability*. Eds. Alison Hope Alkon and Julian Agyeman. Cambridge, MA: MIT Press.

Marx, Karl. 1990 [1867]. *Capital: Volume I*. Translated by Ben Fowkes. London: Penguin Books.

Maye, Damian, Lewis Holloway and Moya Kneafsey. 2007. *Alternative Food Geographies: Representation and Practice*. New York: Elsevier.

McClintock, Nathan. 2010. "Why Farm the City? Theorizing Urban Agriculture through a Lens of Metabolic Rift." *Cambridge Journal of Regions, Economy and Society*, 3: 191–207.

McClintock, Nathan. 2014. "Radical, Reformist, and Garden-variety Neoliberal: Coming to terms with Urban Agriculture's Contradictions." *Local Environment*, 19: 147–171.

McCormack, L.A., M.N. Laska, N.I. Larson et al. 2010. "Review of the Nutritional Implication of Farmers Markets and Community Gardens: A Call for Evaluation and Research Efforts." *Journal of the American Dietetic Association*, 110: 399–408.

McLain, R.J., P.T. Hurley, M.R. Emery and M.R. Poe, 2014. "Gathering 'Wild' Food in the City: Rethinking the Role of Foraging in Urban Ecosystem Planning and Management." *Local Environment*, 19: 220–240. doi:10.1080/13549839.2013.841659.

McMichael, Phillip. 2012. *Development and Social Change: A Global Perspective*, 5th ed. Sage: Los Angeles.

Metcalf, S.S. and Widener, M.J., 2011. "Growing Buffalo's Capacity for Local Food: A Systems Framework for Sustainable Agriculture." *Applied Geography*, 31(4): 1242–1251.

Middlemiss, Lucie. 2018. *Sustainable Consumption*. Philadelphia: Routledge. [VitalSource Bookshelf]. Retrieved from https://bookshelf.vitalsource. com/#/books/9781317239819/.

Mincyte, D. and K. Dobernig. 2016. "Urban Farming in the North American Metropolis: Rethinking Work and Distance in Alternative Agro-Food Networks." *Environment and Planning A*, 48(9): 1767–1786.

Molotch, Harvey. 1976. "The City as a Growth Machine: Toward a Political Economy of Place." *American Journal of Sociology*, 82(2): 309–332.

Moore, L.J. and M. Kosut. 2013. *Buzz: Urban Beekeeping and the Power of the Bee*. New York: New York University Press.

Morales, Alfonso. 2011. "Growing Food and Justice: Dismantling Racism through Sustainable Food Systems." pp. 149–176. In *Cultivating Food Justice: Race, Class and Sustainability*. Eds. Alison Hope Alkon and Julian Agyeman. Cambridge, MA: MIT Press.

National Gardening Association. 2009. "The Impact of Home and Community Gardening in America." Retrieved April 1, 2013 from http://www.gardenresearch.com/files/2009-Impact-of-Gardening-in-America-White-Paper.pdf.

Norgaard, Kari Marie, Ron Reed and Carolina Van Horn. 2011. "A Continuing Legacy: Institutional Racism, Hunger and Nutritional Justice on the Klamath." pp. 23–46. In *Cultivating Food Justice: Race, Class and Sustainability*. Eds. Alison Hope Alkon and Julian Agyeman. Cambridge, MA:MIT Press.

Ober Allen, J., K. Alaimo, D. Elam, et al. 2008. "Growing Vegetables and values: Benefits of Neighborhood-based Community Gardens for Youth Development and nutrition." *Journal of Hunger & Environmental Nutrition*, 3: 418–439.

ORourke, D. and N. Lollo. 2015. "Transforming consumption: From decoupling, to Behavior Change, to System Changes for Sustainable Consumption." *Annual Review of Environment and Resources*, 40: 233–259.

Patel, Rajeev. 2007. *Stuffed and Starved: The Hidden Battle for the World Food System*. New York: Melville House Publishers.

Paxon, Heather. 2013. *The Life of Cheese: Crafting Food and Value in America*. Berkeley, CA: University of California Press.

Pellow, David. 2000. "Environmental Inequality Formation: Toward a Theory of Environmental Injustice." *American Behavioral Scientist*, 43: 581–601.

Polanyi, Karl. 1944. *The Great Transformation: The Political and Economic Origins of Our Time*. Boston, MA: Beacon Press.

Pulighe, G. and F. Lupia. 2016. "Mapping Spatial Patterns of Urban Agriculture in Rome (Italy) using Google Earth and Web-mapping Services." *Land Use Policy*, 59: 49–58. doi:10.1016/j.landusepol.2016.08.001.

Putnam, Robert. 2000. *Bowling Alone: The Collapse and Revival of American Community*. New York: Simon and Schuster.

Roberts, Michael R. 2009. "Control Rights and Capital Structure: An Empirical Investigation." *The Journal of Finance*. https://doi.org/10.1111/j.1540-6261.2009.01476.x.

Russo, A., F.J. Escobedo, G.T. Cirella and S. Zerbe. 2017. "Edible Green Infrastructure: An Approach and Review of Provisioning Ecosystem Services and Disservices in Urban Environments." *Agriculture, Ecosystems & Environment*, 242: 53–66. doi:10.1016/j.agee.2017.03.026.

Saha, M. and M.J. Eckelman. 2017. "Growing Fresh Fruits and Vegetables in an Urban Landscape: A Geospatial Assessment of Ground Level and Rooftop Urban Agriculture Potential in Boston, USA." *Landscape and Urban Planning*, 165: 130–141. doi:10.1016/j.landurbplan.2017.04.015.

Sahakian, Marlene. 2017. "Toward a More Solidaristic Sharing Economy: Examples from Switzerland." in *Social Change and the Coming of Post-Consumer Society*. Philadelphia: Routledge.

Schor, Juliet and Robert Wengronowitz. 2017. "The New Sharing Economy" in *Social Change and the Coming of Post-Consumer Society*. Philadelphia: Routledge.

Sherman, Jennifer. 2009. *Those Who Work, Those Who Don't: Poverty, Morality and Family in Rural America*. Minneapolis: University of Minnesota Press.

Small, Mario L. 2009. "'How many cases do I need?' On Science and the Logic of Case Selection in Field-based Research." *Ethnography*, 10(1): 5–38.

Stack, Carol. 1974. *All our Kin: Strategies for Survival in a Black Community*. New York: Harper and Row.

Teig, E., J. Amulya, L. Bardwell et al. 2009. "Collective Efficacy in Denver, Colorado: Strengthening Neighborhoods and Health through Community Gardens." *Health & Place*, 15, 1115–1122.

UNEP. 2013. "Smallholders, Food Security and the Environment." *International Fund for Agricultural Development*. Retrieved from https://www.ifad.org/documents/10180/666cac24-14b6-43c2-876d-9c2d1f01d5dd.

Urry, John. 2010. "Consuming the Planet to Excess." *Theory, Culture and Society*, 27(2–30): 191–212.

Veen, E.J. 2015. "Community Gardens in Urban Areas: A Critical Reflection on the Extent to which they Strengthen Social Cohesion and Provide Alternative Food." PhD Thesis, Wageningen University, the Netherlands.

Vitiello, Dominic and Laura Wolf-Powers. 2014. "Growing Food to Grow Cities? The Potential of Agriculture for Economic and Community Development in the United States." *Community Development Journal*, 1–16.

Wakefield, S. et al., 2007. "Growing Urban Health: Community Gardening in South-East Toronto." *Health Promotion International*, 22(2): 92–101.

Weber, Max. 1930. *The Protestant Ethic and the Spirit of Capitalism*. Translated by Talcott Parsons. New York: Scribner.

White, Monica M. 2011. "Sisters of the Soil: Urban Gardening as Resistance in Detroit." *Race/Ethnicity: Multidisciplinary Global Contexts. Food Justice*, 5(1): 13–28.

Whiteman, Gail and William H. Cooper. 2000. "Ecological Embeddedness." *Academy of Management Journal*, 43(6): 1265–1282.

Winne, Mark. 2010. *Food Rebels, Guerilla Gardeners, and Smart-Cookin' Mamas: Fighting Back in an Age of Industrial Agriculture*. Boston: Beacon Press.

World Bank. 2008. "World Development Report 2008." Retrieved May 8, 2017 from https://siteresources.worldbank.org/INTWDR2008/Resources/WDR_00_book.pdf.

Wright, Erik Olin. 2010. "Interstitial Transformations." Chapter 10 in *Envisioning Real Utopias*. London: Verso.

York, Richard, Eugene Rosa and Tomas Dietz. 2003. "Footprints on the Earth: The Environmental Consequences of Modernity." *American Sociological Review*, 68(2): 279–300.

2 Subsistence agriculture in South Chicago

The current research takes place in the Southern part of the Chicago metropolitan area, including suburban, rural and urban sites where subsistence food production is taking place. Chicago is a major metropolitan area, the third largest in the United States, with a total population of around 9.5 million (US Census 2017). Chicago is a city notorious for gun violence, although this reputation is somewhat inaccurate as 23 other American cities in fact have higher per capita murder rates (Garbarino 2017). Chicago can in some ways be understood as a part of the former rust belt cities like Detroit and Pittsburgh that lost manufacturing jobs through de-industrialization. This process that started in the last decades of the twentieth century led to the hollowing-out of working-class communities that once thrived in and around the metropolitan area (Bayne 2017). Many of Chicago's employment sectors and neighborhoods remain affluent, but these are in the white-collar industries and the already wealthiest parts of the Chicago metropolitan area (Bayne 2017). The story of Chicago in the context of neoliberal late capitalism is similar to many other American cities after the Great Recession of 2007/2008: growing income inequality and continued crisis for the working class, racial and class-based disparities in social outcomes from health to education to access to transport or adequate food (Grusky, Western and Wimer 2011; Kotlowitz 1992).

What is particularly pronounced in Chicago is the city's famous segregation (Moore 2017), which often has geographic boundaries (e.g. train tracks, major boulevards) that can separate specific ethnic sub-groups. This leads to a kind of separation of Chicago into "two Chicagos": one that has access to resources and one that does not. Journalist Alex Kotlowitz's award winning book *There Are No Children Here* illuminates the life of the "other Chicago" through three years of research on two young African American brothers living in a

Chicago housing project. Kotlowitz found a population within Chicago that echoes the findings of sociologists within disadvantaged populations in many American cities. Lacking access to basic resources such as quality education, food or police protection, these forgotten populations struggle to get by (1992). Kotlowitz argues this is a population who are most negatively impacted by corrupt politics, leaving them without basic human rights, which detracts from their chances for a successful future (1992).

Chicago is also known for its machine politics, which results in the exacerbation of inequalities and access to resources (Royko 1988). This process of neoliberalization and privatization of public resources has only accelerated under the leadership of Mayor Rahm Emmanuel, who came into office in 2011 after long-time famous machine politics mayor Richard Daley retired (Taibbi 2011). Although Chicago is notorious for corruption, the neoliberal model is not unique to this city. In fact, this privatization of public resources is a process that is unfolding across cities in the United States, leading to myriad social problems caused by inequality and lack of access to adequate resources for the most disadvantaged communities (Harvey 2017; Taibbi 2011).

The South Side of Chicago, the setting for the urban field site of this study, is known by social science researchers as a monolith of gang violence, poverty and racial and ethnic segregation. Researchers have shown the way in which natural disasters unequally effect people of color on the south side (Klinenberg 2003), or have taken an in-depth look at south side housing projects which fundamentally represent the American Ghetto (Vankatesh 2002). In fact, the development of criminology came out of an interest by researchers at the University of Chicago in South Side gang activity (Jeffrey 1959; Sampson and Wilson 1995). In this book, however, I make the argument that the South Side of Chicago – alongside the South suburbs and rural communities south of the city – are not one generalizable, monolithic population that is at-risk or impoverished. Instead, I attempt to show a more diverse South Side in terms of race, class, and gender that is adapting, innovating and resilient.

With this field site come particularities that make it not necessarily generalizable to other American cities. However, I chose Chicago with the hope of finding field sites which will help to explain outcomes that are an outgrowth of certain structural conditions, or "the aggregation of situational knowledge into social processes" (Burawoy 1998, 15). This method of sampling has been conceptualized as choosing a synecdoche, which is "a rhetorical figure in which we use a part of something to refer the listener or reader to the whole it belongs to"

(Becker 1998, 67). The theoretical contribution that follows is: when these conditions exist one may find similar outcomes in other places in the United States. This site was chosen due to the easy accessibility of urban, suburban and rural areas in close proximity, and because it is the author's hometown.

I make the case that although Chicago has aspects that make it unique, its similarities to other places experiencing social problems make it a useful field site for the study of emergent shadow structures resulting from the failures of late capitalism (Berman 2017; Harvey 2017). In interpreting the results of this study, I hope to both bear in mind the particularities of Chicago while highlighting the ways in which institutions and social processes that are unfolding here might unfold similarly in other locations with similar sets of characteristics. Chicago is a metropolitan area like many others that is experiencing the unfolding effects of the contradictions of capitalism; especially, the exploitation of labor that ultimately leads to crisis (O'Connor 1991). Because this book is ultimately an exploration of the resultant emerging alternatives to these overlapping crises, I make the case that Chicago is a useful site for understanding this phenomenon and building theory around it.

The (in)significance of geography in the field site

Just like other American cities, the metropolitan Chicago area consists of many diverse and separate places with their own unique histories and socioeconomic contexts. I entered the research with the assumption that geography would make a large impact on the types of SFP that were being practiced. I assumed, for example, that rural communities would have more access to hunting and open spaces for large vegetable gardens and urban communities would have to utilize spaces like community gardens and take part in activities like gleaning fruits from trees planted in public urban spaces. I also entered the research with the assumption that there are certain cultural markers that align with geography such as rural populations accepting hunting and fishing as culturally acceptable (Dean et al. 2016; Sherman 2009) and urban communities being relative newcomers to food growing (Alkon and Agyeman 2011).

The data have disproven my assumptions about the importance of geography on both SFP activities and cultural associations with SFP. Instead, the data suggest that the most salient indicator that determined SFP activities and meanings was class. This will be discussed in more detail in Chapter 4, but it is important to explain here that it was upper-class status that was associated with being a newcomer to SFP activities, and lower-class status that was associated with seeing SFP activity as something people in their family "have always done."

For example, I interviewed upper-class, white, urban couple Chad and Brian, as well as Lance, a wealthy rural white man, who were all very new to SFP, having come to partake in SFP activities such as keeping small livestock or vegetable gardening only within the last decade of their lives. On the other hand, lower-class populations across geography were much more likely to have had a connection to SFP over the course of their family history. For example, Marty, a lower-class white rural man, had been taking part in SFP for his entire lifetime and it was common in his family. Deb, a very low-income woman who identifies as White and Native American, spends part of her time in a dense urban part of Chicago and other parts of the year in a rural cottage. Deb has been producing food alongside her family for the majority of her life. It is class, not geography, that determines the way one approaches SFP.

The range of activities that SFPers took part in was, contrary to my expectations, neither associated with geography nor class. For example, I had SFPers across class, race, and geography taking part in hunting and fishing as well as participants across significant dimensions of difference keeping livestock, gleaning, or keeping large vegetable gardens. It was surprising to find urban participants who fished in the city's park district ponds and Lake Michigan, and to find rural participants taking part in gleaning public fruit.

I entered this research with the plan to sample to maximize geographic range (more detail on sampling strategy is on p. 30) because of what I perceived as a bias toward urban food producers in the literature (Hara et al. 2013; Hardman and Larkham 2014; Pulighe and Lupia 2016; Russo et al. 2017; Saha and Eckelman 2017). I wanted to be able both to include marginalized suburban and urban populations and also to compare across geography. Even though this type of sampling did not produce the differences I expected, I argue that having sampled across geography and including populations that are often marginalized by environmental sociologists has led to important insights about geography and class. Namely, that geography is less a structural limit on SFP activities than the much stronger impact of class.

Despite this finding indicating the insignificance of geographic diversity, it is important for me to situate my work by giving some detail on my field sites. This will provide the reader context through which to understand the population I sought and their social environments. Field sites were categorized into geographical categories by specific markers that generally define urban, rural and suburban. I first used US Census categories as a blunt instrument to identify metropolitan and non-metropolitan areas (OMB 2010). Using this tool, both my urban and suburban field sites are considered a metropolitan area, and my rural field site is considered non-metropolitan.

Researchers like Scott Allard who focus specifically on suburban areas lament the lack of an official tool for categorizing suburban geographies, and often use proxy indicators that reveal a context that is substantively different from dense urban spaces. For Allard's research, he defined suburbs as municipalities that cluster near large metropolitans but are outside of the municipality of the city (2017). In the case of Chicago, this would include areas that are not in the municipality of the City of Chicago, but are adjacent or very close to the municipal border. To categorize my suburban field site, I first used Allard's tool and then further refined this methodology by including markers such as population density, home ownership, and lot size (*Statistical Atlas* 2015).

In the next two sections, I will provide background and context on my urban and suburban field sites together, and the rural section separately. The justification for this is because there is a shared and intertwined history between City of Chicago and the surrounding suburbs that makes their explanation best understood together. The rural field site, however, is quite far removed from the City of Chicago, and has a quite different history and set of social circumstances. For the remainder of the book, I will be using the three separate geographic indicators (urban, suburban and rural) to describe participants. I am using all three indicators instead of two (metro and non-metro) because they provide some social and physical context that can help the data to be situated in a complex set of social circumstances that may help the reader to fully interpret the findings. In the next set of sections, I will describe in some detail how each place's history, especially in relation to subsistence food production, informs the participants' interaction with subsistence food production. I will be using vignettes from the data to help provide some additional depth to the description of the field sites.

Urban and suburban field sites

I met Lucia, a young, working-class Latina woman, for coffee on Chicago's far South Side. She was sweet, laughed a lot, and had a generally upbeat attitude, despite describing a lifelong battle with health issues she attributed to a diet of industrially produced foods.

Within the last few years, fed up with her lack of efficacy making change in her neighborhood through formal political channels, she took over an abandoned lot that had been host to several large industrial factories over the last century or so and turned it into a garden. When she first found this space, it was a typical abandoned urban lot in the hollowed-out rust belt (Bayne 2017): about an acre of concrete, littered with broken glass and trash, surrounded by a chain-link fence, and overgrown with weeds making a life between the cracks in the concrete. When Lucia started asking neighbors about the lot, their stories started to emerge:

> At one point it was a mechanic's shop and trains used to come in there delivering parts and that sort of thing. Then it was owned by a paint company. They used it as a paint and waste processing facility. Dumping a lot of the chemicals directly into the ground. Yeah [laughs] real nice on their part. Eventually, it shut down. The land just sat vacant for a long time, there were still buildings standing there. People started noticing developmental disorders. Right along [the nearest street to the lot]. People were reporting that their kids had a lot of issues. Nobody wanted their kids to go and play there. There were a lot of general illegal activities going on, definitely a lot of drug use.

Lucia described a local history that is similar to many pieces of land that were once part of the industrial production of the now former rust belt states (Bayne 2017). Land that once housed factories that provided jobs for the community; factories that have now closed and left behind toxic, unusable land.

Eventually, Lucia claims that the EPA came in to remediate the land, after some pressure from the surrounding community:

> That horribly toxic land that was just sitting right there in your neighborhood. Got the [EPA] to come in right around the early 2000s. They took off the top two feet of topsoil, put down a clay cap and also some sort of black plastic layer. And then trucked in two new feet of topsoil. Hearing from neighbors it seems that they cut a lot of corners and didn't do it in the safest way possible. They put out a thousand-page report but nobody was able to read it – it was too dense and too long. Unfortunate stuff. It's not surprising but it is unfortunate. Yes, so they put in the new topsoil, which we've since tested and has come out okay. It appears that the clay hasn't been cracking despite putting in new trees.

Lucia describes a process that is repeatedly demonstrated in the environmental justice literature: communities of color or working-class communities are often sites of environmental toxicity, with little power to enact sufficient remediation or obtain justice (Brulle and Pellow 2006; Pellow 2000).

Seeing the abandoned lot as an opportunity to create something positive in their community, Lucia and a few friends petitioned the alderman [city councilperson] to be able to use the space. He gave them a free lease, renewable on a yearly basis. They brought in found materials – an abandoned rowboat, road construction concrete barriers, wood pallets, discarded industrial metal cylinders – filled them with organic soil and turned them into planters. They asked the City of Chicago to bring the woodchips from discarded Christmas trees to their lot, and they inoculated the piles of woodchips with mushroom plugs, with the hopes of pulling toxins from the pile as well as turning them eventually into soil. They tried new methods of food gardening like permaculture and Hügelkultur, swapping information with other farmers and gardeners and food producers they came across. They used the space to invite neighborhood people in. First only kids came, who were amazed to see the process of food growing, but then their parents started coming and sharing seeds and harvesting plants and swapping recipes.

Lucia lives in a community left behind by the promises of the rise of industrialism (Harvey 2017). Once a neighborhood booming with production, it has declined in both population and jobs (Bayne 2017). It is squarely in a food desert (Moser 2012), and racially and economically segregated from wealthier parts of the city (Moore 2017). The people in Lucia's neighborhood, as in many less privileged places around the world, are at higher risk for disparities in health, education, and economic success (D'Abundo and Carden 2008; Freeman et al. 2012; Kremer and DeLiberty 2011; McCormack et al. 2010; Metcalf and Widener 2011; Wakefield et al. 2007).

Small-scale food production is something that has existed in urban and suburban areas in and around Chicago throughout its history. Chicago, like most other American cities, saw an uptick in home food gardens around the time of the Great Depression (Lawson 2005), which were then part of a national effort promoted by the federal government for households to grow "victory gardens." Mirroring the history of home food growing in metropolitan spaces more generally, gardens became less popular in the 1950s and 1960s as the growing middle class began valuing convenience and processed foods increasingly available at supermarkets. New ideas tied up with cultural capital began to see producing one's own food as a negative activity associated with lower class populations and manual labor (Eng 2017; Lawson 2005).

At the same time, within lower-class and especially within waves of immigrant communities, growing one's own was continually practiced as a way to maintain access to culturally relevant food (Gottlieb and Joshi 2010; Mares and Pena 2011). In the last few decades (starting around the time of the Great Recession) home food production has begun to see more widespread increased participation. For example, Mincyte (Mincyte and Dobernig 2016, 1772) argues:

> The United States has witnessed several waves of interest in urban farming during the last century, peaking in times of economic stress or war (Lawson 2005). The most recent uptick in urban farming emerged during and after the 2007/2008 financial crisis.

The results of this book suggest that indeed it is perceived stresses that have drawn the newcomers in my sample into subsistence food production; this is to be discussed in more detail in the Chapter 3.

Chicago has long had ordinances in place that place no limit on the number of chickens a family can keep for egg production, and has never had a ban on keeping roosters, nor any other livestock for food production (American Legal Publishing Corporation 2017, Municipal code 7-12-100). The municipal code on keeping livestock varies by suburban municipality, however. Some suburbs around Chicago have been in a decades-long process of rebranding themselves as sophisticated and modern, and not the rural place they once were (Eng 2017). Within the last decade or so, facing an increased interest in keeping livestock, many suburban communities have overturned newly penned regulations in the face of political resistance (Eng 2017). In fact, the long-held regulation allowing urban livestock in Chicago was challenged in 2007 by a city council member who proposed a city-wide ban. In response, the growing segment of Chicago's chicken keepers mobilized to combat the proposed regulation, which was defeated in the city council (Eng 2017). This incident will be explained in more detail from the perspective of SFPers in my sample in Chapter 6.

The ebbs and flows of urban and suburban Chicago home food production echo the trajectories of many major US metropoli. Interest has shifted with changing political, social and economic circumstances. In the case of my specific urban and suburban field sites, data suggest that perceptions of myriad social problems are pervasive and the drive to take part in subsistence food production is growing. This growing interest in not only home food production but other alternative food network activities is on the rise across the City of Chicago (Block et al. 2011; Eng 2017; Taylor and Lovell 2015) and across other North American cities (Donald et al. 2010; Gottlieb and Joshi 2010; Jarosz 2000; Maye, Holloway and Kneafsey 2007; Morales 2011; White 2011).

Rural field site

I met George, a working-class rural black man who grows most of his own food, at his home on a chilly Spring day. He told me about his concerns for his community, and also for the urban community on Chicago's South Side where he used to live and now sells his produce. The urban population, George feared, was living in an environment that was detrimental to their physical health, "There was no place they could go and eat good. There was no place they could go and stay good. And the patient would come in [to the medical clinic] and get treatment and go back to hell." George created a center out of his rural home for his urban acquaintances to come and experience rural life:

> An hour drive from Chicago and these kids riding around on horses. I just couldn't believe it. They were poor in a certain standard, but they were wealthy rich in nature and horses and open spaces. Just an hour from Chicago! We would get on the highway 45 minutes we were there. Man that was great!

In contrast to Lucia's environment of concrete littered with broken bottles and used drug paraphernalia, George sees the wide-open spaces of his rural region as an important and often overlooked resource. This perception among rural people has been found by other researchers:

> Several rural respondents conveyed that the lack of material wealth is a worthwhile trade-off of living in the country. What they lose in wealth, these respondents feel they gain in the slower pace of life, the spiritual satisfaction, and the simplicity of rural life.
>
> (Kellogg 2002, 4)

George's perception is corroborated by the rural participants in my sample who, despite the challenges, see the value in rural spaces including the ability to have unalienated connection to nature.

Joan is a rural working-class white woman who, with her family, acquires food through SFP by keeping a large vegetable garden, hunting, fishing, canning and foraging. She explains how she learned about gardening, "My mom and dad did it, and their parents, and my aunts and uncles. Well less my uncles than my aunts. And they all fished and hunted, the men did." The perceptions of Joan and George are in some ways unique to their own personal histories, but in other ways are representative of the story of subsistence food production in rural areas more generally. Certainly, it is unlikely that most people are relying solely on food produced through SFP as their principal way to acquire foods.[1] What is known is that SFP is a culturally-accepted practice in this context, and one this population relies on either as a result of recreation or need, depending on their economic circumstance (Dean et al. 2016; Sherman 2009). Further, making a living in rural areas is not easy, and many struggle financially to stay (Kellogg 2002; Laughlin 2016), but do so because they see the benefits of open space, simplicity, and social connections.

My rural field site consisted of two adjoining communities located about an hour's drive outside of the limits of the City of Chicago.[2] The combined population of this area is less than 2,000. One of these communities comprises an overwhelming majority white population, the other majority black, but both are considered working to middle class. The black community has a median income well below the federal poverty line, while the white community has a median income at or slightly exceeding it (US Census 2017). The main sectors where these community members are employed is in the agriculture and service industries, similarly to occupational trends in rural areas across the United States more generally (Laughlin 2016).

For this set of rural communities, as with agricultural communities across North America, farming moved from something practiced by a majority of families for subsistence purposes a century ago to something a small number of families practice at a more industrial scale now (McClintock 2010; VanHaute 2012). The average farm size is over 300 acres and the main crops grown by members of these communities are corn and soybeans, not for self-consumption but usually for either animal feed or further industrial processing (Illinois Department of Agriculture 2018).

Despite gaining significantly less attention by researchers interested in alternative food networks, rural populations generally have taken part in a variety of methods of home food production, which is seen in rural areas as a socially acceptable means of acquiring food in lean economic times (Dean et al. 2016; Sherman 2009). According to reports by working-class participants from this rural field site, the participation in SFP activities, especially hunting, fishing and food gardening, has not had as much change in popularity over the last century as it has in urban areas – it has remained consistently widely practiced. The exception to this is middle-to-upper-class people who have moved to rural areas from urban areas relatively recently. In my sample, these newcomers' trajectory mirrored those of their middle-to-upper class urban counterparts, who are coming to SFP practices only within the last decade or so.

Overall, this rural field site mirrors many of the experiences of rural communities across North America. Left behind by changing economic structures, these rural people struggle to get by. At the same time, they see their rural life as a privilege and a resource, both because of the peace and open space, but also the connection to community it is perceived to provide. Further, producing one's own food for this community, as with most working-class rural communities, is something that is highly practiced and accepted as a fundamental cultural characteristic.

Methods

I conducted 60 semi-structured interviews with subsistence food producers and approximately 120 hours of participant observation mainly at the site of food production for each household, but also at events that attracted the subsistence food production population. I conducted purposive sampling (Mack et al. 2005) to maximize variation in geography, gender, race, and class (see Table 2.1). Field work lasted from February 2015 through May 2016.

For analysis I constructed ideal-type terms of "upper class" and "lower class," with upper class having higher self-reported educational attainment (BA or higher), food secure household,[3] higher-status occupation, and living within an affluent community[4] than lower class. Researchers have found an unsurprising link between food insecurity and lower social class status (Gray et al. 2014; Minkoff-Zern 2014). Of significance to the current research, lower status households have been demonstrated to use alternative means of food acquisition such as home production, hunting and fishing, and bartering as a culturally acceptable way to address food insecurity (Block, Chavez and Allen 2011; Levkoe 2006; Sherman 2009; Winne 2010).

Purposive sampling (Mack et al. 2005) utilizes characteristics of the population as the primary drivers of what kinds of participants are recruited and sampling stops when the researcher reaches theoretical saturation. This is also called "sampling to maximize range" (Weiss 1994, 23) or "treating the full range of cases" (Becker 1998, 107). Many sociological studies that have thus far sought to investigate alternative food networks often look for highly visible organizations (e.g. community gardens) or

Table 2.1 Descriptive statistics of sample

Indicator	N/percentage
Geography[*]	
Urban	23/38%[*]
Suburban	22/37%[*]
Rural	22/37%[*]
Race	
White	33/55%
Nonwhite	27/45%
Class	
Upper/middle	23/38%
Working/lower	37/62%
Gender	
Male	29/48%
Female	31/52%
N = 60	

Notes

[*] Some participants had significant SFP experiences in more than one geographic location. In those cases, those participants where considered both urban and rural, for example, if they both lived and produced food in the city as well as lived part-time and produced food in a rural area.

network through food organizations (Blecha and Leitner 2014; Hara et al. 2013; Hardman and Larkham 2014; McLain et al. 2014; Moore and Kosut 2013; Russo et al. 2017; Saha and Eckelman 2017) to uncover their populations. This leads to a sort of sampling on the dependent variable – producing only organizations or individuals that are associated with *identity* of food producers through association with a group. In the current research, I sampled on self-reported *behavior* – in this case a population that produces at least 50% of food consumption in the high season – independent of membership or connection to highly visible community, often associated with a particular identity. The results reveal a much broader, more diverse, and more inclusive population. Including people who have long self-produced food and those who are new to it, I can compare and contrast meanings and try to understand what ties together disparate individuals to this shared act.

I read through my transcripts from interviews and participant observation and began to develop a coding scheme based on emergent findings using grounded theory (Glaser and Strauss 1967). I first minimized preconceptions and then allowed themes to emerge from data. I began by coding substantive areas, rather than theoretical ones. After this stage, I recorded refined mini-theories (Weiss 1994) that developed from the data already coded. Finally, at the end of the coding process I had developed major findings and theories, which help to explain these findings. I used MaxQDA Qualitative Software to code and analyze my findings, which allows for the entry of demographic characteristics of participants. During the analysis phase, I checked to see if and/or how my codes differed by the different demographic subgroups. I then use characteristic quotations from interviews or description from participant observation, which help to explain the findings from my data.

This book fundamentally tells a story that is both particular and universal. Facing structural crises, and a generalized sense of risk in late industrial capitalism, individuals are turning to subsistence food production as a way to re-connect. What results, as we shall see in the following chapters, is the way in which this seemingly innocuous act can lead to major personal and social transformations. These producers are connecting to work, nature, and community and in the process are working to build new social structures, which I call shadow structures. In the next chapter I will explore the neo-Polanyian lens in which I situate this research, exploring the myriad crises being faced in the failure of global capitalism, and the interesting, surprising, and innovative activities that are arising as a result.

Notes

1 Although no comprehensive cross-national surveys have been conducted to confirm or deny the extent to which rural populations rely on SFP practices for food access.
2 Because the anonymity of participants in a rural setting with a population of less than 2,000 is harder to protect solely with changes in participant name, for my rural field site I will be changing minor details and place names to protect those who took part in the study.
3 Based on the short-form, six question USDA ERS Food Security Survey Module, as asked within the semi-structured interview (USDA ERS 2012).
4 Area median income >200% of national median (US Census 2010, American Community Survey 2009–2013)

Works cited

Alkon, Alison Hope and Julian Agyeman. 2011. "Introduction: The Food Movement as a Polyculture." pp. 1–20. In *Cultivating Food Justice: Race, Class and Sustainability*. Eds. Alison Hope Alkon and Julian Agyeman. Cambridge, MA: MIT Press.
Allard, Scott. 2017. *Places in Need: The Changing Geography of Poverty*. New York: Russell Sage Foundation.

American Community Survey 2009–2013. "Median Income." Retrieved May 18, 2020 https://www.census.gov/data/developers/updates/acs-5-yr-summary-available-2009-2013.html.

American Legal Publishing Corporation. 2017. "Chicago Municipal Code 7-12-100." Retrieved May 8, 2018 from http://library.amlegal.com/nxt/gateway.dll/Illinois/chicago_il/municipalcodeofchicago?f=templates$fn=default.htm$3.0$vid=amlegal:chicago_il.

Bayne, Martha. 2017. *Rust Belt Chicago: An Anthology*. Chicago: Belt Publishing.

Becker, Howard S. 1998. *Tricks of the Trade: How to Think About your Research While you're Doing It*. Chicago: University of Chicago Press.

Berman, Morris. 2017. "Dual Process: The Only Game in Town." In *Are We There Yet?* Brattleboro, VT: Echo Point Books. Essay #27.

Blecha, Jennifer and Helga Leitner. 2014. "Reimagining the Food System, the Economy, and Urban Life: New Urban Chicken-keepers in US Cities." *Urban Geography*, 35(1): 86–108.

Block, Daniel R., Noel Chavez and Erika Allen. 2011. "Food Sovereignty, Urban Food Access, and Food Activism: Contemplating the Connections through Examples from Chicago." *Agriculture and Human Values*, 29(2): 203–215.

Brulle, Robert and David Pellow. 2006. "Environmental Justice: Human Health and Environmental Inequalities." *Annual Review of Public Health*, 27: 103–124.

Burawoy, Michael. 1998. "The Extended Case Method." *Sociological Theory*, 16(1): 4–33.

D'Abundo, M. and A. Carden. 2008. "Growing Wellness: The Possibility of Collective Wellness through Community Garden Education Programs." *Community Development*, 39: 83–94.

Dean, Wesley R., Joseph R. Sharkey and Cassandra M. Johnson. 2016. "The Possibilities and Limitations of Personal Agency: The Walmart that Got Away and Other Narratives of Food Acquisition in Rural Texas." *Food, Culture and Society*, 19(1): 129–149.

Donald, B. et al. 2010. "Re-regionalizing the Food System?" *Cambridge Journal of Regions, Economy and Society*, 3(2): 171–175.

Eng, Monica. 2017. "Chickens and goats and pigs oh my! Chicago's Backyard Livestock Laws." *WBEZ: Curious City*. Retrieved June 2, 2018 from https://www.wbez.org/shows/curious-city/chickens-and-goats-and-pigs-oh-my-chicagos-backyard-livestock-laws/b08ac437-8d53-4bc5-8130-565d84f5c1e6.

Freeman, C. et al. 2012. "My garden is an expression of me": Exploring Householders' Relationships with their Gardens." *Journal of Environmental Psychology*, 32(2): 135–143.

Garbarino, James. 2017. "Gun Violence in Chicago." *Violence and Gender*, 4(2): 45–47.

Glaser, Barney and Anselm Strauss. 1967. *The Discovery of Grounded Theory*. Chicago: Aldine.

Gottlieb, R. and Joshi, A., 2010. *Food Justice.* Cambridge, MA: MIT Press.

Gray, L. et al. 2014. "Can Home Gardens Scale Up into Movements for Social Change? The Role of Home Gardens in Providing Food Security and Community Change in San Jose, California." *Local Environment,* 19(2): 187–203.

Grusky, David B., Bruce Western and Christoper Wimer (eds.). 2011. *The Great Recession.* New York: Russell Sage.

Hara, Y., A. Murakami, K. Tsuchiya, A.M. Palijon and M. Yokohari. 2013. "A Quantitative Assessment of Vegetable Farming on Vacant Lots in an Urban Fringe Area in Metro Manila: Can it Sustain Long-term Local Vegetable Demand?" *Applied Geography,* 41, 195–206. doi.org/10.1016/j.apgeog.2013.04.003.

Hardman, Michael and Peter Larkham. 2014. *Informal Urban Agriculture: The Secret Lives of Guerilla Gardeners.* New York: Springer International Publishing.

Harvey, David. 2017. *Marx, Capital, and the Madness of Economic Reason.* New York: Oxford University Press.

Illinois Department of Agriculture. 2018. "Facts about Illinois Agriculture." Retrieved June 3, 2018 from https://www2.illinois.gov/sites/agr/About/Pages/Facts-About-Illinois-Agriculture.aspx#h2.

Jarosz, L. 2000. "Understanding Agri-food Networks as Social Relations." *Agriculture and Human Values,* 17(3): 279–283.

Jeffrey, Clarence Ray. 1959. "The Historical Development of Criminology." *Journal of Criminal Law and Criminology,* 50(1).

Kellogg, W.K. Foundation. 2002. "Perceptions of Rural America." Retrieved June 3, 2018 from https://www.wkkf.org/resource-directory/resource/2002/12/perceptions-of-rural-america.

Klinenberg, Eric. 2003. *Heat Wave: A Social Autopsy of Disaster.* Chicago: University of Chicago Press.

Kotlowitz, Alex. 1992. *There Are No Children Here: The Story of Two Boys Growing up in the Other America.* Chicago: Doubleday.

Kremer, P. and T.L. DeLiberty. 2011. "Local Food Practices and Growing Potential: Mapping the Case of Philadelphia." *Applied Geography,* 31(4): 1252–1261.

Laughlin, Linda. 2016. "Beyond the Farm: Rural Industry Workers in America." *Census Blogs.* Retrieved June 3, 2018 from https://www.census.gov/newsroom/blogs/random-samplings/2016/12/beyond_the_farm_rur.html.

Lawson, Laura. 2005. *City Bountiful: A Century of Community Gardening in America.* Berkeley, CA: University of California Press.

Levkoe, C.Z. 2006. "Learning Democracy through Food Justice Movements." *Agriculture and Human Values,* 23: 89–98.

Mack, Natasha, Cynthia Woodsong, Kathleen Macqueen, Greg Guest and Emily Namey. 2005. *Qualitative Research Methods: A Data Collector's Field Guide.* Retrieved November 15, 2013 from http://www.ccs.neu.edu/course/is4800sp12/resources/qualmethods.pdf.

Mares, Teresa M. and Devon G. Pena. 2011. "Environmental and Food Justice: Toward Local, Slow and Deep Food Systems." pp. 197–219. In *Cultivating Food Justice: Race, Class and Sustainability*. Eds. Alison Hope Alkon and Julian Agyeman. Cambridge, MA: MIT Press.

Maye, Damian, Lewis Holloway and Moya Kneafsey. 2007. *Alternative Food Geographies: Representation and Practice*. New York: Elsevier.

McCormack, L.A., M.N. Laska, N.I. Larson et al. 2010. "Review of the Nutritional Implication of Farmers Markets and Community Gardens: A Call for Evaluation and Research Efforts." *Journal of the American Dietetic Association*, 110: 399–408.

McLain, R.J., P.T. Hurley, M.R. Emery and M.R. Poe. 2014. "Gathering 'Wild' Food in the City: Rethinking the Role of Foraging in Urban Ecosystem Planning and Management." *Local Environment*, 19: 220–240. doi:10.1080/13549839.2013.841659.

Metcalf, S.S. and M.J. Widener. 2011. "Growing Buffalo's Capacity for Local Food: A Systems Framework for Sustainable Agriculture." *Applied Geography*, 31(4): 1242–1251.

Mincyte, D. and K. Dobernig. 2016. "Urban Farming in the North American Metropolis: Rethinking Work and Distance in Alternative Agro-Food Networks." *Environment and Planning A*, 48(9): 1767–1786.

Minkoff-Zern, L-A., 2014. "Hunger Amidst Plenty: Farmworker Food Insecurity and Coping Strategies in California." *Local Environment*, 19(2): 204–219.

Moore, L.J. and M. Kosut. 2013. *Buzz: Urban Beekeeping and the Power of the Bee*. New York: New York University Press.

Moore, Natalie. 2017. *The South Side: A Portrait of American Segregation*. Chicago: Picador.

Morales, Alfonso. 2011. "Growing Food and Justice: Dismantling Racism through Sustainable Food Systems." pp. 149–176. In *Cultivating Food Justice: Race, Class and Sustainability*. Eds. Alison Hope Alkon and Julian Agyeman. Cambridge, MA: MIT Press.

Moser, Whet. 2012. "Challenges to the concept and effect of the food desert." *Chicago Magazine*. Retrieved June 3, 2018 from http://www.chicagomag.com/Chicago-Magazine/The-312/April-2012/Challenges-to-the-Concept-and-Effects-of-the-Food-Desert/.

O'Connor, James. 1991. "On the Two Contradictions of Capitalism." *Capitalism, Nature, Socialism*, 2(3): 107–109.

OMB (Office of Management and Budget). 2010. "2010 Standards for Delineating Metropolitan and Micropolitan Statistical Areas." Retrieved June 3, 2018 from https://www.gpo.gov/fdsys/pkg/FR-2010-06-28/pdf/2010-15605.pdf.

Pellow, David. 2000. "Environmental Inequality Formation: Toward a Theory of Environmental Injustice." *American Behavioral Scientist*, 43: 581–601.

Pulighe, G., Lupia, F., 2016. "Mapping Spatial Patterns of Urban Agriculture in Rome (Italy) using Google Earth and Web-mapping Services." *Land Use Policy*, 59: 49–58. doi:10.1016/j.landusepol.2016.08.001.

Royko, Mike. 1988. *Boss: Richard J. Daley of Chicago*. Chicago: Plume.

Russo, A., F.J. Escobedo, G.T. Cirella and S. Zerbe. 2017. "Edible Green Infrastructure: An Approach and Review of Provisioning Ecosystem Services and Disservices in Urban Environments." *Agriculture, Ecosystems & Environment*, 242: 53–66. doi:10.1016/j.agee.2017.03.026.

Saha, M. and M.J. Eckelman. 2017. "Growing Fresh Fruits and Vegetables in an Urban Landscape: A Geospatial Assessment of Ground Level and Rooftop Urban Agriculture Potential in Boston, USA." *Landscape and Urban Planning*, 165: 130–141. doi:10.1016/j.landurbplan.2017.04.015.

Sampson, Robert J. and William Julius Wilson. 1995. "Toward a Theory of Race, Crime, and Urban Inequality." In *Crime and Inequality*, eds John Hagan and Ruth D. Peterson. Stanford, CA: Stanford University Press.

Sherman, Jennifer. 2009. *Those Who Work, Those Who Don't: Poverty, Morality and Family in Rural America*. Minneapolis: University of Minnesota Press.

Statistical Atlas. 2015. "Race and Ethnicity Beverly, Chicago, Illinois." Retrieved May 15, 2018 from https://statisticalatlas.com/neighborhood/Illinois/Chicago/Beverly/Race-and-Ethnicity.

Taibbi, Matt. 2011. *Griftopia: A Story of Bankers, Politicians, and the Most Audacious Power Grab in American History*. New York: Spiegel and Grau.

Taylor, J.R. and S.T. Lovell. 2012. "Mapping Public and Private Spaces of Urban Agriculture in Chicago through the Analysis of High-resolution Aerial Images in Google Earth." *Landscape and Urban Planning*, 108(1): 57–70.

USDA ERS. 2012. "Food Security Survey Module." Retrieved May 18, 2020 from https://www.ers.usda.gov/media/8282/short2012.pdf.

US Census. 2017. "U.S. Gazeteer Files." Retrieved June 3, 2018 from https://www.census.gov/geo/maps-data/data/gazetteer.html.

VanHaute, E. 2012. *Handbook of World-Systems Analysis*. Retrieved February 12, 2017 from http://www.ccc.ugent.be/file/116.

Venkatesh, Sudhir. 2002. *American Project: The Rise and Fall of a Modern Ghetto*. Cambridge, MA: Harvard University Press.

Wakefield, S. et al. 2007. "Growing Urban Health: Community Gardening in South-East Toronto." *Health Promotion International*, 22(2): 92–101.

Weiss, Robert S. 1994. *Learning from Strangers: The Art and Method of Qualitative Interview*. New York: The Free Press.

White, Monica M. 2011. "Sisters of the Soil: Urban Gardening as Resistance in Detroit." *Race/Ethnicity: Multidisciplinary Global Contexts. Food Justice*, 5(1): 13–28.

Winne, Mark. 2010. *Food Rebels, Guerilla Gardeners, and Smart-Cookin' Mamas: Fighting Back in an Age of Industrial Agriculture*. Boston: Beacon Press.

3   Guiding theories

Social problems, emergent solutions

> We learn from our gardens to deal with the most urgent question of
> the time:
> How much is enough?
>
> – Wendell Berry (1977)

There are three guiding theories on which I rely on for the remainder
of this book: 1) I first explain in detail the historical sociological
theory of Dual Process, and the resultant *shadow structures*, which is
a term I use to characterize the group of phenomena that subsistence
food production (SFP) belongs in, theoretically (Berman 2017).
2) I then use the modern, environmental re-interpretations of
Polanyi, Marx and Weber's work on the effects of alienation and
disenchantment in modern, industrial life and how both the original
and modern scholars understand the phenomenon of *countering* that
alienation among certain populations (Berman 1981; Foster 1999;
Foster and Holleman 2012). 3) Finally, I draw on the work of recent
theoretical scholarship of Galt, Grey and Hurley (2014), McClintock
(2014), and Wright (2010) that argues that the development of
shadow structures will be subject to complex forces that will result in
dialectical, contradictory and sometimes paradoxical outcomes, and
the importance of accepting these complexities as important in the
overall understanding of these phenomena.

Dual process and shadow structures

Adding to the findings of the discipline of environmental sociology
documenting myriad environmental social problems, scholars across
disciplines have also recognized other crises unfolding in the global

social, economic and political realms (Beling et al. 2017; Fischer-Kowalski and Haberl 2007). These overlapping catastrophes may lead to another inevitable "Great Transformation" (Polanyi 1944) – akin to the industrial revolution – either "by design or by disaster" (Leonard 2011). Prominent sociologist John Urry suggests, "the twentieth century has left a bleak legacy for the new century, with a very limited range of possible future scenarios" (2010, 191). This book accepts this as an assumption of the research, but seeks to explore the resilience, innovation and creativity that is happening within this "limited range."

Following the idea that global capitalist society is facing a fundamental transformation in the 21st century, some scholars have focused on how alternatives to the logic of mainstream capitalism are emerging (Berman 2017; McClintock 2014; Beck 2016). Ulrich Beck is the sociologist famous for his understanding of modern society as one replete with inherent risks (Beck and Ritter 1992), and posthumously released innovative research suggests global society is not just in a state of change, but in metamorphosis:

> Metamorphosis implies a much more radical transformation in which the old certainties of modern society are falling away and something quite new is emerging. To grasp this metamorphosis of the world it is necessary to explore the new beginnings, to focus on what is emerging from the old and seek to grasp future structures and norms in the turmoil of the present.
>
> (Beck 2016, 1)

To understand the *process* of this impending (or currently unfolding) metamorphosis, I draw heavily on historian Morris Berman's neo-Polanyian theory of dual process to explain the context in which my population is situated (2017).

There are two important aspects to dual process theory: 1) the slow collapse of modern capitalism (that Berman argues runs from 1500 CE through 2100 CE), and 2) the alternatives that then emerge as a response to that collapse. People turn to alternatives, Berman argues, because the reigning socioeconomic system is no longer meeting their basic needs, and as these new activities grow and adapt they may ultimately replace the hegemonic system.

Embedded here are two important characteristics of Berman's theory that do not exist in most thinking on systemic alternatives. First, Berman, in the tradition of Marx and Polanyi, takes as a fundamental assumption that capitalism is collapsing and will end due to its fundamental contradictions and exploitations. Just as the Roman Empire transformed into the Christian Empire, and feudalism transformed into capitalism, we are in an era of history where there are structural and physical limits that portend the end of the current economic and social system. This inevitability impacts the way in which he considers alternatives. Instead of engaging in *prescriptive* ideas about how things *should* be done to usher in a new era (as is the case for much of the literature of sustainable consumption), instead he is engaging in *descriptive* work that explains *what is already happening* through this theoretical lens.

Second, Berman argues that these alternatives rarely seek to attack or confront the entirety of the capitalist system directly, instead they are like "shadow economic, social, and even technological structure [s] that will be ready to take over as the existing system fails" (Ehrenfeld in Berman 2017, 1). That is, the work individuals are doing to develop shadow structures may or may not be informed by the hope or explicit intention of overthrowing capitalism. Instead, shadow structures develop out of a desire meaningfully to solve practical problems people face as capitalism fails to provide for their needs. As I demonstrate throughout this book, focusing on individuals exploring practical alternatives leads to a more inclusive, diverse community, rather than one whose criteria for inclusion is some sort of ideology. The current research focuses therefore not on the crises emergent in the era of late capitalism, but on the second part of dual process: the resilient alternatives emerging as a result of these crises.

In order to refer back to this concept with more clarity and brevity, for the remainder of this book I will use Ehrenfeld's term *"shadow structures,"*[1] which can be any sort of individual or group activity that adopts an alternative logic to that of late capitalism. I argue that this type of activity can be thought of as a sort of Polanyian double movement, discussed in the introduction, in which individuals push back against their sense of the extreme commodification of land, labor and money (Polanyi 1944). Individuals exploring shadow structures are attempting to create small-scale social structures (economic, political, cultural) that meet presented needs and exist parallel to (not in direct conflict with) the hegemonic industrial system. Some examples of shadow structures include small-scale local food production, alternative banking schemes, gift or sharing economies like tool or seed libraries, decentralized currency, co-operative housing, community energy production, and others discussed in the introduction.

An earlier mention of dual process by theologian Jeorg Rieger in

2006 describes in slightly more detail how the development of alternatives to late capitalism may develop, with some examples:

> First starve the capitalist system through demythologization of its ideology and clear refusal (strikes, boycotts of banks and cheap consumer goods) and resistance. The second step includes developing local-regional alternatives to capitalism (exchange-based rather than money-based; cooperative banking, alternative energies, local food production), reclaiming resources in all areas (dignified work, agriculture, public goods such as water, just taxes–capital pays less and less, social networks, fair trade, renewable energies), and developing an alternative macro-narrative of hope.
>
> (1)

Here Rieger suggests as the larger system fails, alternative institutions such as local food production can begin to develop in earnest alongside the collapse.

Very recent work by prominent author and activist Naomi Klein found evidence in Puerto Rico of shadow structures in the form of organic farms and food co-ops that existed as the economic and infrastructure systems were failing, especially after devastating Hurricane Maria. The part of the population that was involved in these local food operations had access to food as the rest of the island was without power and reliant upon imported food to return (Klein 2018). Another well-known example of a shadow structure is Cuba's organic agriculture movement. When faced with an embargo of outside resources – that is, when locked out of the hegemonic global economic system – Cubans experimented with low-input sustainable agriculture and were extremely successful in providing enough high-quality food for the island's residents (Rossett 2002).

In his work, Berman draws attention to the fact that his process is playing out in places in which the contradictions of capitalism are already occurring (Harvey 2017), and where the policies of late capitalism lead to populations unable to meet basic needs. For example, countries implementing economic austerity measures in the wake of the Great Recession of 2007–2008 (e.g. Greece, Spain, Portugal, Ireland) are especially rife with emergent alternative economic institutions such as cooperatives (joint or democratic ownership of an enterprise) or time banks (exchanging services without currency). In Spain there are at least 325 burgeoning alternative currencies, and in Barcelona alone there are 100 time banks (Berman 2017). A number of nation states are already interested in alternative economic policies: like de-growth, which is "an equitable downscaling of production and consumption that increases human well-being and enhances ecological conditions" (Schneider, Kallis, and Martinez-Alier 2010, 511). An example of upending standard economic indicators by inventing new measurements is Growth National Happiness used in Bhutan rather than Gross Domestic Product (OPHI 2017).

Berman argues for the importance of looking at grassroots, small-scale or individual actions to best understand these alternatives, "The folks who have the real answer to this austerity mess, which is a crisis of the entire neoliberal model, are not in the government. They are the folks at the grass-roots level, pursuing a form of Dual Process" (2017, 1). These individual actions lead to a development of alternatives that exist parallel to the mainstream systems until a point at which the larger organizations more systematically fail and the alternatives then emerge to take their place (Berman 2017).[2] Although I will be focusing exclusively on what is often called alternative food networks (McClintock 2014) as one example of emergent alternatives, the burgeoning literature on this topic may include: sharing, non-monetary, or gift economies (Schor and Wengronowitz 2016; Sahakian 2017, Transition Towns (Hopkins 2014), alternative currencies and time banks (Berman 2017; Lietaer 2001; Lietaer et al 2012), alternative modes of transportation, downshifters (Kennedy, Krahn and Krogman 2013), voluntary simplicity, freecycling or dumpster diving (Ferrell 2008), among others, explored in some detail in the introduction.

This book is fundamentally about investigating the *process* of shadow structure development, and is part of a larger set of research looking at alternatives to the failing social structures of late capitalism. Very few researchers have contributed to the study of these emergent shadow structures while engaging with central sociological theory. My research moves beyond the work that has already been done in this area, which tends to be highly descriptive and has a strong urban bias, by adding a layer of theoretical understanding to the development of shadow structures. In the next section, I will highlight in some detail how I will use modern reinterpretations of classical theory (Foster 1999; Foster and Holleman 2012; West 1985) to understand how the action of SFP is tied up some of the central questions within the social sciences, which include feelings of alienation in the modern era, and the meanings and responses to such a sensation.

Countering alienation

The foundation of sociology was laid by scholars looking to understand the social problems emerging from the onset of the modern era. Marx is most famous for his exploration of capitalist exploitation of the worker and the resultant feelings of *alienation* (1990). Weber focused on the outcomes on social structures in the modern era, such as the changing relationships in society from Gemeinschaft to Gesellschaft, and the resultant feelings of *disenchantment* (1930). Modern scholars have focused in a more holistic way in which the founders' work can be reinterpreted to understand the myriad ways in which industrial society has fundamentally alienated individuals from work, nature and community (Berman 1981; Berman 2017; Foster 1999; Foster and Holleman 2012; King 2008; Mincyte and Dobernig 2016).

Max Weber is known for his analyses of culture (1930) as well as for the importance placed on rationalization, bureaucracy, and the state in the formation of inequality (Collins 1994). Central to Weberian environmentalism is the way in which culture influences different societies' relationship with the natural world (West 1985). It is important to note however, that this relationship between culture and nature is decidedly *not* a deterministic one, but cultural ideas about how to interact with the environment are constantly competing and evolving, which leads to "selective survival over time" (233). Ideas are especially important as they can "end up becoming forces in themselves" (Foster and Holleman 2012, 633).

One way in which Weber puts this theory to use is in the explanation of the move from the pre-modern (traditional/organic) to the modern (rational/inorganic) era. Weber wrote a number of volumes describing this difference between modern and pre-modern Western society (see Table 3.1 below; adapted from Foster and Holleman 2012). The findings of this book suggest that there is a significant resurgence of interest in concepts that help to counter the alienation of modernity.

Table 3.1 Weber's characteristics of pre-modern and modern eras

Traditional/organic	*Rational/inorganic*
Organic limitations to production	Increasingly growing production with agricultural chemistry
Business in home	Business separate from home
Common land	Private land/enclosure
Closed nutrient cycle	Linear nutrient cycle
Enchantment	Disenchantment/rationalization/alienation
Machine subservience to men	Men subservient to machines
Gemeinschaft/subjective community ties	Gesellschaft/indirect, rational social ties

Note
Adapted from the work of Foster and Holleman 2012.

The drive to fight against the angst of alienation can come through connection to manual work (explored in Chapter 4), connection to nature (Chapter 5), or connection to others (Chapter 6). Just briefly treated here (see Table 3.1), Weber characterizes the ideas of these two eras, and how they heavily influence the way in which society uses resources and treats waste.

This schema can be adapted for use in the study of subsistence food production. Data suggest that the drive to take part in SFP is motivated by a desire to return to some of the qualities of the traditional/organic era. If, following Weber's argument, ideas can mediate one's relationship with the natural world, I argue that research attempting to understand the ideas of those pursuing shadow structure development is important. For example, in Chapter 5, I explore the ways in which SFPers are driven by their sense of connection to the natural world to explore closed nutrient cycles, organic limitations to production, and conservation of natural resources through SFP. In Chapter 6, I make sense of the ways in which subjective community ties (Granovetter 1983) of *Gemeinschaft* are an outgrowth of the decision to take part in SFP. Weber's theory of disenchantment and re-enchantment, of alienation and connection, helps us to understand the ways in which modern individuals are making sense of these large-scale processes, and how those meanings mediate their relationship with the Earth (1930).

Where Weber focuses on how cultural meanings intercede in our interaction with the environment, Marx is more interested in how the capitalist economic system mediates that relationship. Further, as Weber divides time into the pre-modern and modern eras, Marx thinks in terms of dialectical forces over time. Although Marxist environmental sociologists have mostly utilized metabolic rift theory put forth by John Bellamy Foster (1999), I argue that there is much more that can be used in Marx's work to help understand modern responses to environmental problems.

Central to Marxist environmental theory is the relation of human societies to nature through agriculture. Marx argued that with industrial agriculture came the widespread use of external chemical inputs (both fertilizer and energy for farm machines) as well as the linearization of what was once a cyclical flow of materials (Marx 1981). This concept is defined by Foster as the metabolic rift (1999), or the disturbance of the natural cycle of nutrients (also discussed as linear nutrient cycle by Weber, see Table 3.1 above). The ecological problems of capitalist agriculture were the focus of Marx's early environmental critiques:

> Large-scale industry and industrially pursued large-scale agriculture have the same effect...The former lays waste and ruins the

labour-power and thus the natural power of man, whereas the latter does the same to the natural power of the soil.

(Marx 1981, 950)

These critiques evolved into larger concerns with both agriculture and industry, the exploitation of both the worker and the land, and the resultant feelings of alienation (Foster 1999).

Subsistence production can fundamentally change production relations and may also create a concept of wealth that includes nature (Foster 1999, 388). According to Foster, Marx argued that industrial agriculture is fundamentally not sustainable, and "demanded a radical transformation of the human relation to the earth via changed production relations" (Foster 1999, 386). The part of Marxist environmental theory that most closely maps onto the study of SFP is the concept of alienation, or "the material estrangement of human beings in capitalist society from the natural conditions of their existence" (Foster 1999, 383). Alienation from both the material conditions of one existence, the means of production as well as from nature are themes that continue to arise throughout the remainder of this book, explored through the lens of food production.

It may be that food is not simply another commodity to be bought and sold coldly in the global marketplace. Instead, food has meaning in many cultures around the world. Taking away that fundamental connection of people with nature – through their food sources – may explain why this specific issue has gotten people from very disparate cultures to agree that industrial agriculture removes this connection and is not the way forward. Prominent geographer Phillip McMichael puts it more succinctly:

> Food is not just an item of consumption, it's actually a way of life. It has deep material and symbolic power. And because it embodies the links between nature, human survival and health, culture and livelihood, it will, and has already, become a focus of contention and resistance to a corporate takeover of life itself.
>
> (2012, 31–32)

The data suggest that subsistence food production is a way in which individuals are reconnecting with nature and labor through agriculture, and in doing so they are attempting to counter feelings of alienation.

Marx and Weber's original writings and their modern re-interpretations can be used for the study of sustainable responses to the environmental problems brought on by modernity. Both classical theorists place great emphasis on the importance of the changes in agriculture and the disastrous social and ecological consequences of those changes in the transition to modernity. Whereas Weber places more weight on cultural meanings in determining how societies interact with the natural world, Marx cares more about economic structures.

In the study of shadow structures, I argue that the meanings brought to SFP are central to understanding the ways in which cultural change in rejection of industrial capitalism may be occurring. In fact, many aspects of subsistence food production reported by participants draw on phenomena commonly present in the traditional/organic era (Table 3.1). Utilizing the work of Foster and Holleman (2012) and Foster (1999), I find that the meanings brought to subsistence production indeed result from some of the same feelings of alienation, disenchantment and commodification described by the forefathers of sociology: Marx, Weber, and Polanyi. In Chapter 4, I explore the way in which meanings associated with *countering the alienation* of modernity cuts across social boundaries of race, class and geography. Further, in Chapter 5 I find that the desire to connect with nature through the act of SFP is another way in which participants describe attempting to *reenchant* their daily lives (Berman 1981). In Chapter 6 I find that subsistence producers are pursuing a Gemeinschaft-like community structure that upends the logic of social relations in the era of late capitalism.

Paradox in shadow structure development

Finally, I draw on the work of geographers McClintock (2014), Galt et al. (2014), and sociologist Wright (2010) to understand how this overall alienation and response to the collapse of the institutions of late capitalism leads to the development of alternatives, but that this development is not a clean, clear or linear process. Instead, as with most social phenomena, this process is messy, complex and full of contradictions. In this section, I will explore the ways in which food growing has been demonstrated to be related to larger socioeconomic trends, and how the act of self-production has been critiqued as neoliberal. Then I will draw on researchers' suggestions for ways in which we can study SFP and other shadow structures in order to understand them more fully.

Theorists studying these alternatives suggest we consider the possibility that this process is dialectical and can be rife with contradictions (McClintock 2014; Galt et al. 2014; Wright 2010). For example, on the one hand, individualized solutions to societal issues represent a type of neoliberal logic, which can be defined as "a theory of political economic practices that proposes that human well-being can best be advanced by liberating individual entrepreneurial freedoms and skills" (Harvey 2005, 2). In other words, neoliberal logic puts an emphasis on economic growth, often through the success of corporations, as the greatest means of providing freedom to all the members of society. Social issues, according to this logic, are best solved either individually or through the free market. This logic was present in my study when producers like Carter and Lydia who described their decision to take part in producing food as "think globally, act locally," which in some ways represents a logic of privatizing social solutions. Yet, in other ways, subsistence production represents a radical break from capitalist logic in taking individual control over the means of production, and producing meaningful and pragmatic solutions to perceived problems (Kennedy, Johnston and Parkins 2017).

Growing food for household consumption is an act that has often been tied up with economic need in the face of social-historical phenomena (Lawson 2005), as will be discussed in some detail in the next chapter. Yet, many researchers have critiqued this reliance on private citizens to meet their own needs as advancing neoliberal logic by putting the onus on individuals to provide their own social safety nets in the face of mounting economic uncertainty (Guthman 2008), especially among low-income communities where services have already been significantly reduced. Understood this way, the process of growing one's own food does, in some important ways, sanction the logic of individualized solutions to public issues.

Yet, at the same time, this process can also be understood in more complex terms, as geographers Galt et al. (2014) argue. Positive aspects of individual food production include "increases in community and individual initiative, self-sufficiency, and self-help" (141). These authors suggest we move beyond the categories of either/or; as in, individual food production is either good or it is bad. Instead, they suggest a "both/and strategy" (Galt et al. 2014, 141) that recognizes *both* the contributions of those taking part in this phenomenon, *and* engages in the larger discourse that calls into question who is responsible for taking care of basic human needs. These researchers are arguing that the outcome of taking part in the development of alternative food networks (and, I argue, in the development of shadow structures more generally) can be at once a reflection of neoliberal values as well as a radical break from late capitalist logic. I agree that as researchers of these phenomena we must move beyond dualistic categories of bad/good, neoliberal/radical, and so on to be able to describe the complex and paradoxical ways in which these phenomena can be both/and as well as ever evolving in dialectical tension.

Galt et al. (2014) go on to lament how this kind of dualistic cate-
gorization of social solutions is inherently grounded in the politics of
modern liberalism. They argue that:

> Solving every problem at once is usually not possible through
> everyday practice. There is a common paralysis among the traditional
> left in which oppositional politics leads to the constant deferral of
> non-capitalist/transformative initiatives since they are seen as standing
> little chance to actually be structurally transformative (Gibson-
> Graham 2006). The problems of these initiatives should not be
> glossed over, but they should also not be used as imperfections that
> justify the dismissal of these actually existing alternatives...dismissal
> in this way can be a manifestation of a *debilitating politics of perfection*
> that will not create widespread change, and that can hide and
> perpetuate hegemony (cf. Gibson-Graham 2006; DuPuis et al. 2011).
> (Galt et al. 2014, 135, emphasis added)

Here Galt and colleagues (2014) suggest that academic researchers
should not follow what they describe as the pitfalls of the traditional
left that utilizes dualistic categories in which a solution is either fully
acceptable (and often politically unfeasible) or fully unacceptable
because it contains unacceptable elements. Further, as researchers we
should be careful not to engage too much in a *prescriptive* politics,
describing what we think *ought to* happen. Instead, researchers should
investigate the ways in which these processes are actually developing,
and can be *both* neoliberal *and* radical, conservative and progressive,
or hegemonic and transformative.

Sociologist Erik Olin Wright's theory of Interstitial Transformation
(2010) explores the creative processes that get developed in the cracks and
spaces of the dominant system of power. For Wright, the amount of work
put into interstitial transformation *before* capitalism inevitably collapses,
the smoother the transition to a new system will be. The development of
these interstitial strategies will be challenged by capitalist structures, at
which time individuals or communities will develop new interstitial strategies that can overcome these limits. We will see examples of this in
Chapter 6, where the community of chicken keepers' ability to keep
chickens was threatened, and how they organized as a result. According to
Wright, "there will thus be a kind of cycle of extension of social empowerment and stagnation as successive limits are encountered and eroded"
(2010, 235). At the end of this long process, Wright envisions one scenario
in which enough development of and engagement in alternatives that the
state will be infiltrated and subsequently transformed into a socialist state.

Another scenario Wright suggests is possible is one in which interstitial strategies are thwarted by capitalist (or state-authoritarian) structures, and as capitalism fails more "ruptural" strategies will be put into place to un-block the limits placed on developing more interstitial processes.

Geographer Nathan McClintock's work focuses on the development of one specific set of shadow structures – alternative food networks – in the era of late capitalism. McClintock draws on Marxist theory in suggesting that alternative food networks are in a constant dialectical tension between the world of neoliberalism in which they exist, the resultant radical counter-movement away from that logic, and the interplay between the two (2014, 148). McClintock suggests that it is indeed the logic of capit-alism and its many contradictory processes that "both create opportunities for urban agriculture and impose obstacles to its expansion. Identifying these contradictions requires analysis of urban agriculture's various forms and functions at multiple scales" (2014, 148).

I will explore these contradictions in both the literature and the world it seeks to describe. For example, in Chapter 4 I discuss the ways in which typically class-based meanings around consumption have been shown to strengthen social boundaries, while other connotations have the potential to unite across dimensions of social difference. I consider in Chapter 6 the way in which the decision to take part in SFP is often tied up with privatized solutions that exemplify neoliberal logic, yet surprisingly radical social connections and resultant shadow structures grow out of the need to reach out to others for help in SFP practices. Indeed, I argue that food is a social material that is particularly subject to the contradictions of human societies. As the feminist geographers Allison and Jessica Hayes-Conroy note, the meanings ascribed to food can be simultaneously good/bad, junky/healthy, neoliberal/radical. Eating, she argues, "like all human action – is imperfect and contra-dictory" (2008, 362). If we can accept the difficulties of embracing these contradictions, researchers will be able to understand the phenomenon as a complex set of processes. "Such a view is characterized by neither flinty-eyed realism nor dewy-eyed romanticism but rather by a creative synthesis of the two" (Palmer in Galt et al. 2014, 131).

I hold this paradox in mind throughout the remainder of this book, and hope to make the case that future researchers in this area accept the possibility of contradictory findings as acceptable and even plausible as we move forward to better understand the emergent phenomena of shadow structures more generally. In this book I explore contradictions among this particular sub-movement of shadow structures in the face of the challenges of late capitalism. Subsistence food producers are responding to the challenges of their specific social moment with creativity and resilience in the face of difficult challenges. This response is part of a *process* and rife with socially specific *meanings*, something exploratory qualitative research is exceptional at uncovering (Small 2009). I attempt to show a glimpse into a specific group looking to be resilient in the face of potentially transformative social, political, economic and ecological crises.

Notes

1 The creation of this term draws heavily on the words of Ehrenfeld and Berman (2016) in Dual Process theory. Within the development of Dual Process theory, however, this term has not been used explicitly to succinctly describe the emergent alternatives of the dual process. I have therefore decided to coin this term "shadow structure" to more easily refer to this idea. I contacted Ehrenfeld directly to inquire about the primary source of this quotation, and he acknowledged that the quote was his but did not know its published source (direct communication 2017). I acknowledge that the term is not mine, it is heavily reliant upon the ideas of these scholars.
2 I will discuss the tension between private or individual level of action and public sphere action in more detail in Chapter 6.

Works cited

Beck, Ulrich. 2016. *The Metamorphosis of the World: How Climate Change is Transforming our Concept of the World.* New York: Polity Press.

Beck, U. and Ritter, M. 1992. *Risk Society: Towards a New Modernity.* London: Sage Publications.

Beling, Adrian, Julien Vanhulst, Federico Dermaria and Jerome Pelenc. 2017. "Discursive Synergies for a 'Great Transformation' towards Sustainability: Pragmatic Contributions to a Necessary Dialogue between Human Development, Degrowth and Buen Vivir." *Ecological Economics.* doi:10.1016/j.ecolecon.2017.08.025.

Berman, Morris. 1981. *The Reenchantment of the World.* Ithaca, NY: Cornell University Press.

Berman, Morris. 2017. "Dual Process: The Only Game in Town." In *Are We There Yet?* Brattleboro, VT: Echo Point Books. Essay #27.

Berry, Wendell. 1977. *The Unsettling of America: Culture and Agriculture.* San Francisco: Sierra Club Books.

Collins, Randall. 1994. *Four Sociological Traditions.* New York: Oxford University Press.

Ferrell, Jeff. 2008. "Happiness is a Warm Dumpster." Core Connections Lecture Series. University of New England. Biddeford, Maine.

Fischer-Kowalski, M. and H. Haberl. 2007. *Socioecological Transitions and Social Change: Trajectories of Social Metabolism and Land Use.* In "Advances in Ecological Economics," series ed. Jeroen van den Bergh. Cheltenham, UK and Northampton, USA: Edward Elgar.

Foster, John Bellamy. 1999. "Marx's Theory of Metabolic Rift: Classical Foundations for Environmental Sociology." *The American Journal of Sociology* 105(2): 366–405.

Foster, John Bellamy and Hannah Holleman. 2012. "Weber and the Environment: Classical Foundations for a Post-exemptionalist Sociology." *American Journal of Sociology* 117(6): 1625–1673.

Galt, Ryan E., Leslie C. Gray, and Patrick Hurley. 2014. "Subversive and Interstitial Food Spaces: Transforming Selves, Societies, and Society–Environment Relations through Urban Agriculture and Foraging." *Local Environment* 19(2): 133–146.

Gibson-Graham, J.K. 2006. *A Postcapitalist Politics.* Minneapolis: University of Minnesota Press.

Granovetter, Mark. 1983. "The Strength of Weak Ties." pp. 201–233. In *Sociological Theory.* Volume 1.

Guthman, Julie. 2008. "Neoliberalism and the Making of Food Politics in California." *Geoforum* 39(3): 1171–1183.

Harvey, David. 2005. *A Brief History of Neoliberalism.* Oxford University Press: New York.

Harvey, David. 2017. *Marx, Capital, and the Madness of Economic Reason.* New York: Oxford University Press.

Hayes-Conroy, Allison and Jessica Hayes-Conroy. 2008. "Taking Back Taste: Feminism, Food and Visceral Politics." *Gender, Place, and Culture* 15(5): 461–473.

Hopkins, Rob. 2014. *The Transition Handbook: From Oil Dependency to Local Resilience.* UIT Cambridge Limited.

Kennedy, E.H., H. Krahn, and N.T. Krogman. 2013. "Downshifting: An Exploration of Motivations, Quality of Life, and Environmental Practices." *Sociological Forum* 28(4): 764–783.

Kennedy, E.H., J. Johnston and J. Parkins. 2017. "Small-p politics: How Pleasurable, Convivial, and Pragmatic Political Ideals Influence Engagement in Eat-Local Initiatives." *British Journal of Sociology.* doi:10.1111/1468-4446.12298.

King, Christine A. 2008. "Community Resilience and Contemporary Agri-Ecological Systems: Reconnecting People and Food, and People with People." *Systems Research and Behavioral Science* 25: 111–124.

Klein, Naomi. 2018. "Puerto Ricans and Ultrarich 'Puertopeans' are Locked in a Pitched Struggle over How to Remake the Island." *The Intercept.* Retrieved May 17, 2018 from https://theintercept.com/2018/03/20/puerto-rico-hurricane-maria-recovery/.

Lawson, Laura. 2005. *City Bountiful: A Century of Community Gardening in America.* Berkeley, CA: University of California Press.

Leonard, Annie. 2011. "Global Change: By Disaster or by Design?" Presentation at Tulane University, New Orleans, LA. October 3, 2011.

Lietaer, B., Arnsperger, C., Goerner S. et al. 2012. "Money and Sustainability: The Missing Link." Axminster: Club of Rome – EU Chapter.

Lietaer, B. 2001. *The Future of Money: A New Way to Create Wealth, Work and a Wiser World*. London: Random House.

Marx, Karl. 1981 [1863–1865]. *Capital: Volume III*. New York: Vintage.

Marx, Karl. 1990 [1867]. *Capital: Volume I*. Trans. Ben Fowkes. London: Penguin Books.

McClintock, Nathan. 2014. "Radical, Reformist, and Garden-Variety Neoliberal: Coming to Terms with Urban Agriculture's Contradictions." *Local Environment* 19: 147–171.

McMichael, Phillip. 2012. *Development and Social Change: A Global Perspective*, 5th ed. Los Angeles: Sage.

Mincyte, D. and K. Dobernig. 2016. "Urban Farming in the North American Metropolis: Rethinking Work and Distance in Alternative Agro-Food Networks," *Environment and Planning A* 48(9): 1767–1786.

OPHI. 2017. "Bhutan's Gross National Happiness Index." Retrieved May 15, 2018 from http://www.ophi.org.uk/policy/national-policy/gross-national-happiness-index/.

Polanyi, Karl. 1944. *The Great Transformation: The Political and Economic Origins of Our Time*. Boston, MA: Beacon Press.

Rossett, Peter. 2002. *The Greening of the Revolution: Cuba's Experiment with Organic Agriculture*. New York: Ocean Press.

Rieger, Jeorg. 2006. "That's Not Fair': Upside-Down Justice in the Midst of Empire." pp. 91–106. In *Interpreting the Postmodern: Responses to "Radical Orthodoxy."* Edited by Rosemary Radford Ruether and Marion Grau. New York: T&T Clark International, 2006.

Sahakian, Marlyne. 2017. "Toward a more Solidaristic Sharing Economy: Examples from Switzerland." in *Social Change and the Coming of Post-Consumer Society*. New York: Taylor and Francis.

Schneider, F., G. Kallis and J. Martinez-Alier. 2010. "Crisis or Opportunity? Economic Degrowth for Social Equity and Ecological Sustainability." *Journal of Cleaner Production*, 18(6): 511–518.

Schor, Juliet and Robert Wengronowitz. 2016. "The New Sharing Economy." In *Social Change and the Coming of Post-Consumer Society*. Philadelphia: Routledge.

Small, Mario L. 2010. "'How Many Cases Do I Need?' On Science and the Logic of Case Selection in Field-Based Research." *Ethnography*, 10(1): 5–38.

Urry, John. 2009. "Consuming the planet to excess." *Theory, Culture and Society*, 27(2–30): 191–212.

Weber, Max. 1930. *The Protestant Ethic and the Spirit of Capitalism*. Translated by Talcott Parsons. New York: Scribner.

West, Patrick C. 1985. "Max Weber's Human Ecology of Historical Societies." pp. 216–243. In *Theory of Liberty, Legitimacy and Power: New Directions in the Intellectual and Scientific Legacy of Max Weber*. Boston: Routledge & Kegan Paul.

Wright, Erik Olin. 2010. "Interstitial Transformations." Chapter 10 in *Envisioning Real Utopias*. London: Verso.

4   Who are subsistence food producers in Chicago?

Meanings across class of alienation and viscerality

Sustainable consumption scholars have long been interested in the interaction between class and consumption, with most scholarship tending to focus on wealthier demographics who have the social capital to choose to engage in proposed changes in consumption patterns (Boli and Thomas 1997; Boucher 2017; Frank, Hironaka and Shofer 2000; Franzen and Mayer 2010; Inglehart 1995; Middlemiss 2018). Keeping in mind how power and complex politics impact different social groups' ability to participate in certain initiatives (Isenhour, Martiskainen and Middlemiss 2019), it is imperative to explore aspects of sustainable consumption that have the potential to be inclusive and diverse, while exploring marginalized voices.

This study differs from most other explorations of sustainable consumption in two important ways. First, this is an exploration of an act of *production*, rather than changes in patterns of *consumption*. This, as we shall see throughout the book, leads to significantly lower barriers to entry as well as different outcomes for potential social change. Second, this study was designed to include diversity on dimensions of difference such as race, class, geography, and age. Instead of specifically choosing to develop the study around a homophilous group, like high cultural capital environmentalist consumers, this study is interested in exploring individuals that represent dimensions of social difference, with the commonality of shared behavior.

In this chapter, I explore how certain meanings such as viscerality (Hayes-Conroy 2011), ethical consumption (Beagan, Chapman and Power 2016; Kennedy, Johnston and Parkins 2017) and countering alienation (Foster and Holleman 2012; Mincyte and Dobernig 2016) cut across class[1] boundaries and unify subsistence food producers, with important implications. The findings contribute to the extensive literature on class theory (Bourdieu 1984; Lamont 1992) by using modern interpretations of classical sociological theory (Foster and Holleman 2012) to explain the ways in which the meanings around SFP can bring together diverse populations (Galt, Gray and Hurley 2014; McClintock 2014).

Paradox, shared meanings (alienation and viscerality)

Most recent studies on how food is deployed as a form of cultural capital have focused on *consumption* rather than the *production* of food (e.g. Carfagna et al. 2014; Haluza-Delay 2008; Kasper 2009; Micheletti and Stolle 2005). In the study of food tastes and consumption practices, sociologists have found a set of meanings that generally cluster around two categories: high cultural capital and low cultural capital. It is important to first note that it has been demonstrated through survey data that high cultural capital consumption practices have been correlated with high education and/or income (Elliott 2013; Micheletti and Stolle 2005; Neilson and Paxton 2010; Newman and Bartels 2011; Willis and Schor 2012), thereby providing evidence for the ways in which consumption meanings and orientations can align with and reinforce significant socioeconomic indicators.

Despite the seemingly distinct categories described in this body of research that set high cultural capital and low cultural capital food orientations apart from one another, recent scholarship has begun to understand the complex ways in which these class boundaries are beginning to shift and blend. Following the work of McClintock (2014) and Galt and colleagues (2014) who suggest the study of alternative food systems must move beyond dualistic categories (e.g., high versus low), I will explore literature here that complicates these findings in the realm of food meanings and cultural boundaries.

In the introduction, I argued for the importance of the work of Polanyi, Marx and Weber in understanding the transition to modernity. These scholars argue that the modern era has fundamentally changed individuals' relationship to work, nature and each other. The transition to modernity has mostly been in the direction of commodification (Burawoy 2017; Harvey 2017; Polanyi 1944), alienation (Foster 1999; Harvey 2017; Marx 1990) and disenchantment (Berman 1981; Foster and Holleman 2012; Weber 1930) from these fundamental aspects of human life. More recent scholarship has engaged with these foundational texts in order to explain why populations are attempting to counter alienation either from manual work or exposure to natural conditions. Mincyte and Dobernig (2016), for example, explore the ways in which relatively privileged populations successfully choose to take part in farming experiences as a way of directly experiencing non-alienated work and ecosystems.

Recent scholarship has also focused on the ways in which manual work through craftsmanship is accessible across social boundaries (Campbell 2005; Crawford 2009; Ocejo 2017) and puts the means of production into the hands of individuals taking part in these activities, and thereby re-embeds meaning in work. Studies focusing on countering alienation in food systems help to explain the shortening of distance through embeddedness. Bowen (2011) uses localized and horizontal systems of production and exchange of cheese in France to display a way in which food production and consumption can counter alienation and foster a sense of place and space. Other scholars have looked specifically at the ways in which participation in community garden food production can help to re-embed individuals in work, natural systems and community across social class (Kurtz 2001; Teig et al. 2009; Veen 2015).

The study done by Kennedy, Baumann and Johnston (forthcoming) suggests that an orientation toward ethical consumption, or the desire to consume products with social and environmental issues in mind, is not necessarily only a high cultural capital phenomenon. Indeed, the desire to consume ethically produced, quality, healthy foods cuts across class. This finding has been demonstrated repeatedly: those lower socioeconomic status populations that hold ethical consumer orientations are often limited by economics (Beagan, Chapman and Power 2016; Nevarez, Tobin and Waltermauer 2016; Nielsen and Holm 2016). Recent research by Schoolman (2017) continues to complicate the "foodie" and "ethical consumer" tastes as only belonging to high SES consumers in his finding that low cultural capital consumers value the ethics of local consumption, driven by the desire to promote local businesses. All of these studies upend the notion of a working-class consumer driven simply by the rudimentary need to fulfill caloric requirements (Maguire 2017). Instead, the tastes, preferences and motivations of low SES consumers work in complex and creative ways to upend traditional notions that only those with high cultural capital have refined tastes around food.

Other studies of food developed in the field of geography emphasize the viscerality of food and its embodiedness that can impact the way it is perceived and consumed within different subcultures. Viscerality in this work can be defined as "the realm of internally-felt sensations, moods and states of being, which are born from sensory engagement with the material world" (Hayes-Conroy and Hayes-Conroy 2008, 462). In other words, the term viscerality used in the study of SFP is the physicality inherent in the production, processing and consumption of food.

The data from the current study suggest, alongside the findings of several scholars in geography (Goodman 2015; Hayes-Conroy 2010; McMichael 2012), that it is indeed the experience of food's physicality, and the connection to manual work in its production, that can allow it to move between and among more dualistic boundaries such as upper and lower classes. The viscerality of food can, in some cases, work to upend cultural or class-based distinctions simply due to the shared experience all humans have of preparing and eating. The data suggest that the meanings surrounding the *production* of food have an even stronger ability to cut across class boundaries. I argue this is partly due to the physical nature of SFP, and the shared experience of that embodiedness of manual work that is very closely tied up with feelings of countering alienation within the population.

Other scholars critique this understanding of food viscerality, claiming that political and social policies that influence the ability to access food, for example, are just one way in which food is deeply embedded in social institutions and that boundaries are built into those institutions (Goodman 2015). That is, the physical nature of food is not enough to counteract the ways in which social institutions work to exercise power and allocate privilege to different populations. The current study found elements of both food production crossing class distinctions as well as strengthening boundaries. For the purposes of this chapter, I will be focusing primarily on the ways in which certain meanings traversed boundaries of social difference. Nearly every participant in my sample described a sense of alienation from food, nature, or physical work and a desire to counter alienation through the act of SFP. Similarly to scholars' recent interpretation of Polanyi, Marx and Weber's work, a large component of SFP is a reaction to the commodification, alienation or disenchantment of industrialism and capitalism (Berman 1981; Foster 1999; Foster and Holleman 2012; Harvey 2017), and these feelings find a home in the viscerality of food (Hayes-Conroy 2011).

Thus far in this book, I have explored a set of neo-Polanyian theories (Berman 2017; McClintock 2014; Galt et al. 2014; Wright 2010) that are interested in the ways in which individuals are exploring alternatives to the failing logic of industrial capitalism. Scholars of sustainable consumption have primarily focused on limitations and class differences in sustainable consumption patterns. In this section, I discuss the ways in which, surprisingly, subsistence food producers are developing meanings that cut across distinctions of not only class, but also race, gender, age and geography. This finding adds some detail to the mostly theoretical neo-Polanyian literature, by exploring the developing, complex meanings that result from individuals exploring interstitial spaces (Wright 2010) by re-embedding in work, nature and community to counter the alienation brought on by industrialization and late capitalism (Foster 1999; Foster and Holleman 2012; Mincyte and Dobernig 2016).

Alienation

The meanings that cut across these social distinctions are two-sided. On the one hand is a fundamental distrust in the quality of food being produced today and the myriad potential consequences of continually consuming that food. The underlying theme is a sense of alienation from food, work, and community, as well as a deep understanding of the processes that are necessary for survival. The response to that distrust or skepticism is the desire to counter alienation and find a solution to the problem of food quality through work and self-production of food. There are two paradoxical outcomes to this response. In one sense, SFP leads to a kind of re-enchantment with both natural processes and the work involved in food production. Yet, this solution can also buttress the neoliberal logic of personalized or privatized solutions to problems of the public sphere. This finding suggests a deep distrust in systems of governance in producing adequate solutions, which will be discussed in more detail in Chapter 6.

Lucia is a lower-class Latina woman who has a large vegetable garden and also practices a combination of gleaning, bartering, food preservation, and has a large mushroom cultivation project. Here she describes the process of coming to SFP, and what motivated her decision to start producing food:

> [I have had] a lifelong struggle with health issues. It is really the basis for what inspired me to do it in the first place...I can't really eat very much meat because of antibiotics. All things that *are a result of the current food system* that we have. I was very medicalized for a very long time, and nobody really knew what to do with this relatively new [diet-related] disease...I started learning all these [diet-based] tactics as a sort of practical response to all of these diseases that I had and that I saw other people having. And, you know, the high concentrations in Chicago and my family [of diet-based diseases]. I didn't want to go back to traditional, or Western medicine. Especially because they didn't even know what to do with me.

Lucia is describing a sense of danger from industrial processes and this fundamental distrust can be described as a sort of process of disenchantment (Berman 1981; Foster and Holleman 2012; Weber 1930) and risk (Beck 2016; Beck and Ritter 1992) in feeling unable to rely on industrially produced food for health. That is, Lucia sees the food available to her through supermarkets as profoundly dangerous to her overall health and wellbeing. In this way, she is distrustful of the basic systems put in place by her society to provide her sustenance and is in the process of searching for alternate means of acquiring what she perceives to be healthy food as well as healing her body.

George, a lower-class black man living in a rural area mentioned earlier, describes his similar process of coming to SFP:

> I didn't know anything about nothing. I worked in [the auto] industry. I didn't care about no environment. I didn't care about no food. I didn't know anything about that crap. And [a solar energy instructor I met] said 'You've gotta read this guy. This guy is incredible.' Richard Heinberg. *The Party's Over*. And I read that book and it messed me up. I was so messed up after reading that. I went downtown and did that exercise. It said imagine what it would be like to have 10% less energy. I went on the lakefront and I looked at the skyline and I said 'oh my god. This is horrible.' I never had no background in anything. So it started as a journey for my whole family.

Here George was exposed to concepts related to peak oil, energy decline, and environmental problems. This started him on a process of learning more about both environmental problems and solutions to those problems. This process was one of disenchantment and alienation (Berman 1981; Foster 1999; Foster and Holleman 2012; Mincyte and Dobernig 2016) from the system he was relying on for safety and security, producing a fundamental feeling of distrust and dread, and resulted in George exploring alternate means of providing food, shelter and energy for himself, his family and his community.

Carter, an upper-class suburban white man mentioned earlier, describes why he came to SFP, "I think people are very distrustful of the food system. I mean what is organic? What does that mean? People don't trust it. It has become sort of a corporate selling brand as opposed to a meaningful declaration of food quality." Carter is describing a deep sense of suspicion in government labeling of foods. This skepticism in the neoliberal logic in accurate food labeling has driven Carter to seek alternate means of acquiring food where the processes involved in the production are both known to him and within his control.

Most participants in the sample also suggested a sense of strain from everyday existence in a consumer capitalist society. Ignacio, a lower-class Latino man living in a dense urban area who grows vegetables, keeps chickens, gleans, barters and forages, suggested:

> With many things in life it's hard to measure that you're making progress on things because everything is so nebulous and vague and you're like "Am I really, like being successful in this facet of life?"

Here Ignacio describes a generalized sense of meaninglessness or lack of clarity throughout aspects of his "normal," routine life. Throughout my sample, participants suggested reaching a breaking point wherein the strain of the malaise or alienation reached a point in which they felt they had to try something new, or explore a kind of solution outside of their normal purview.

It is important to note that the perceived specific cause of the social issue varied greatly among my participants. However, nearly every participant mentioned becoming aware of what they described as the strain from large social issues, which led to their choice to self-produce food. Social issues mentioned by participants include: environmental and energy problems (as evidenced by George), large-scale health issues (Lucia), distrust of industrial products (Carter), or even awareness of the psychological strain of everyday existence (Ignacio). Despite the diversity of causes, they are nearly all tied to some aspect of modernization and industrialization and the myriad social problems associated with this historical period. Further, they represent a fundamental distrust in the system of production and consumption that is typically available in the hegemonic capitalist system (Foster and Holleman 2012; Weber 1930; West 1985). The response to this perceived alienation may be a form of Marxian dialectic in which these individuals are driven by the shortcomings of the capitalist system to imagine alternatives to it (Foster 1999; Marx 1990; Mincyte and Dobernig 2016; Wright 2010).

Viscerality, re-enchantment

Not only did the majority of participants mention some sense of distrust and disaffection from the system on which they rely to meet their basic needs, they also consistently reported the response to this perceived alienation impacted their choice to take part in SFP. The data suggest that respondents perceived food production as somehow a more visceral (Hayes-Conroy 2011), or embodied, experience through physical work and connection to nature. Danny, a lower-class white urban man, describes the desire to go hunting and fishing as motivated by getting out of the rut of his everyday existence. He describes a particular memory from hunting:

> [The hunting dog] runs down the field, and then he comes back whatever direction the wind is blowing and he will have his nose up in the air and all of a sudden he will stop and get closer and closer. And when he gets closer, he'll point. It's one of the [laughs] coolest things you ever will see. It's something you have to experience.

This kind of joy in connecting with nature (Hayes-Conroy 2011), and the emphasis on the physical, experiential component of it, described by the SFPers in my sample was nearly universal.

Mary and Liz, a lower-class white couple living in an urban setting, have a large vegetable garden and chickens. Here Liz described her idea of the benefits of SFP:

> I liked being outside just in general. So, maybe I had an appreciation for just being outside and just my observation of what was around me, the trees. We grew up on a little pond, too, so I'd see birds and we'd go fishing and I'd see fish and dragon flies or bumble bees flying around, and maybe the migration for different birds, things like that.

Ignacio describes a similar experience:

> When you are outdoors in the garden you've probably been expending a lot of energy, and then you've got this beautiful scenery and the sensory stuff going on with the wind and the sunshine and all of that...with gardening it's like one of those things where you can: plant seed, add water, grow produce, and you know it's happening. It's also one of those things where you can practice mindfulness. Where you can really be involved in the moment. It's really a way to simplify life, and focus on what we're doing here.

In Ignacio's language we can see a couple of important themes. Similarly to Danny and Liz, there is a sensory, physical (Hayes-Conroy 2011) aspect of SFP that is important and meaningful to him, and it is something that draws him to this behavior – especially in the connection to and appreciation of natural elements (something that will be discussed in more detail in Chapter 5).

More than that, there is a sense that this kind of activity is more tangible, more real, more measurable, and less alienated than other aspects of his daily life. This second part of Ignacio's comment echoes a Marxist sentiment wherein:

> Unalienated work constitutes the center of human life and the source of self-respect, value, and meaning, such an approach suggests that the rift between food consumers and agricultural producers—and more broadly, between the country and the city, and nature and industry—can be resolved by qualitatively changing the relations of production and consumption.
>
> (Mincyte and Dobernig 2016, 1768)

According to Marx (and modern re-interpretations), there are parallel processes of alienation from work, nature and community that result from industrial society. Here, through the act of fundamentally changing food production, Ignacio is reporting a sense of connection to work and nature through the experience of growing food in his garden.

Lydia, an upper-class white woman mentioned earlier in the chapter, describes an analogous experience:

> I work hard on my spiritual connection to the earth. I see her [the Earth] as a living being who has essentially fed us every meal we've ever eaten. And we are so disconnected from that fact. So, saying grace before you eat, but for me it's really remembering where it came from, and for me it's from the Earth, our home. I think people intellectualize it and feel guilty. But to really tap into the miracle of that. There's something very – it connects me to the earth. I go out every day. I look at the garden, I look at the sky, I look at the trees, I watch the cycle of the seasons happen because I am caring for these farm animals. There's this whole other perspective about life that you just don't get. I mean, you can intellectualize it and you could read books. But unless you are going out into that garden and looking at it each day, there's something very meaningful and rich and true. It's like you're participating in life.

Here Lydia draws a distinction between intellectual experiences and the more visceral practice of being outdoors and participating in SFP. For her, there is an important difference between even thinking, reading and talking about SFP and participating in it experientially. Through the act of SFP, Lydia has found important meaning, and feels, like Ignacio, that it is in some ways a more true or real activity than others in her life.

Despite diversity in my sample of SFP in terms of race, class, gender and geography, nearly every participant mentioned a sense of disenchantment or alienation (Berman 1981; Foster 1999; Foster and Holleman 2012; Marx 1990 [1867]; Weber 1930) and the resultant response of choosing to take part in SFP as a way to counter that alienation (Mincyte and Dobernig 2016). The data suggest that SFP is perceived as a more visceral (Hayes-Conroy 2011) or more "real" activity than they normally get to experience in their lives. This finding adds research to the recently built theory of environmental sociologists who have re-interpreted Marx and Weber to understand the ways in which the perception of life in the industrial (or now late capitalist) era leads to a sense of commodification, disenchantment, alienation (Berman 1981 Burawoy 2017; Foster 1999; Foster and Holleman 2012) from work, nature and community. The theory suggests that this sense of alienation eventually leads to a sort of dialectic response, a double movement (Polanyi 1944), that has the potential to drive individuals to shorten the distance between themselves and their food by taking control of the means of production.

This finding relies on the theory of Hayes-Conroy (2011) that the visceral, or physical/material, nature of food can in some instances blur some social boundaries. The perception that working outdoors with food in nature is a more real experience cuts across important dimensions of difference within my sample. This finding adds to the literature much-needed nuance and understanding of moving beyond dualistic categories of high and low consumers (Maguire 2017), and finds evidence for Hayes-Conroy's claim that "articulated relationships to food cannot possibly be one thing – whether 'bad' or 'good', 'neoliberal' or 'anti-capitalist', 'junky' or 'healthy'" (2011, 74). The immediate implication of this finding is the suggestion that meanings around connection, viscerality, and countering alienation that cut across social boundaries may be a jumping off point for more social connection and potential political action. I will build on these findings to explore the ways in which the desire to connect to nature cuts across environmentalist identity (Chapter 5) and the social and political implications of these shared meanings and behaviors (Chapter 6).

Embracing paradoxical findings

The current research suggests that much of how subsistence food production emerges and the meanings associated with it are paradoxical, and we must be able to accept these paradoxical results. This is in line with the seminal piece on the study of alternative food networks by McClintock, who chastises the political left for relying too heavily on a "politics of perfection" in which if some aspect of a cultural change or social movement is dismissed if it is imperfect in some way. Instead, McClintock (2014) argues that the emergence of alternatives arising alongside the unfolding crises of late capitalism will be rife with paradox, and that accepting and understanding these paradoxes helps researchers to move beyond all-or-nothing or unhelpful dualistic categories.

Past research has found significant ways in which social classes emphasize symbolic boundaries through enacting cultural capital, and how power and politics shape one's ability to participate in movements of sustainable consumption (Isenhour, Martiskainen and Middlemiss 2019). High cultural capital consumers orient toward "foodie" culture which values authenticity, exoticism and quality (Kennedy et al. in press). Low cultural capital consumers tend to focus on family foods and "tastes of [economic] necessity" (Baumann, Szabo and Johnston 2017; Bourdieu 1984). However, these distinctions begin to blur when we find many shared meanings that cut across class including the feelings that food produced through SFP is enacting a sort of physicality (Hayes-Conroy 2011) that counters alienation and helps to address issues of ethical consumption that most SFPers perceive.

McClintock's (2014) framework for embracing paradox allows us to upend distinctions through shared meanings centered around countering alienation. If we take up the call from Maguire (2017), Hayes-Conroy (2011), McClintock (2014), and others to move beyond the dualistic categories of high versus low consumers, we can find the ways in which shared meanings can have potentially transformative social outcomes. The data suggest that this desire to counter the commodification of all aspects of life is part of a neo-Polanyian dialectic (Berman 2017; Galt et al. 2014; McClintock 2014; Wright 2010) in which segments of the population have reached a point of distrust of the hegemonic system and alienation from meaningful connection to work, nature and community that they have turned to SFP to help to counter this trend.

Note

1 For a more detailed discussion of how I define class, please see Chapter 2, section "Methods."

Works cited

Baumann, Shyon, Michelle Szabo and Josee Johnston. 2017. "Understanding the Food Preferences of People of Low Socioeconomic Status." *Journal of Consumer Culture*. doi:10.1177/1469540517717780.

Beagan, Brenda L., Gwen E. Chapman and Elaine M. Power. 2016. "Cultural and Symbolic Capital with and without Economic Constraint Food Shopping in Low-income and High-income Canadian Families." *Food, Culture & Society*, 19(1): 45–70.

Beck, Ulrich. 2016. *The Metamorphosis of the World: How Climate Change is Transforming our Concept of the World*. New York: Polity Press.

Beck, U. and M. Ritter. 1992. *Risk Society: Towards a New Modernity*. London: Sage Publications.

Berman, Morris. 2017. "Dual Process: The Only Game in Town." in *Are We There Yet?* Brattleboro, VT: Echo Point Books. Essay #27.

Berman, Morris. 1981. The Reenchantment of the World. Ithaca, NY: Cornell University Press.

Boli, John and George M. Thomas. 1997. "World Culture in the World Polity: A Century of International Non-Governmental Organization." *American Sociological Review*, 62(2): 171–190.

Boucher, J.L. 2017. "Culture, Carbon and Climate Change: A Class Analysis of Climate Change Belief, Lifestyle Lock-in, and Personal Carbon Footprint." Socijalna Ekologija, 25(1): 53–80.

Bourdieu, Pierre. 1984. *Distinction: A Social Critique of The Judgement of Taste*. Cambridge, MA: Harvard University Press.

Bowen, S. 2011. "The Importance of Place: Reterritorialising Embeddedness." *Sociologia Ruralis*, 51: 325–348.

Burawoy, Michael. 2017. "Social Movements in the Neoliberal Age." pp. 21–35. In *Southern Resistance in Critical Perspective*, eds. M. Paret, C. Runciman, L. Sinwell. New York: Routledge.

Campbell, Colin. 2005. "The Craft Consumer: Culture, Craft and Consumption in a Postmodern Society." *The Journal of Consumer Culture*, 5(1): 23–42.

Carfagna, Lindsey B., Emilie A. Dubois, Conner Fitzmaurice, Monique Ouimette, Juliet B. Schor, Margaret Willis and Thomas Laidley. 2014. "An Emerging Eco-habitus: The Reconfiguration of High Cultural Capital Practices among Ethical Consumers." *Journal of Consumer Culture*, 14(2): 158–178.

Crawford, Matthew B. 2009. *Shop Class as Soulcraft: An Inquiry into the Value of Work*. New York: Penguin.

Elliott, Rebecca. 2013. "The Taste for Green: The Possibilities and Dynamics of Status Differentiation through 'Green' Consumption." *Poetics*, 41(3): 294–322.

Foster, John Bellamy. 1999. "Marx's Theory of Metabolic Rift: Classical Foundations for Environmental Sociology." *The American Journal of Sociology*, 105(2): 366–405.

Foster, John Bellamy and Hannah Holleman. 2012. "Weber and the Environment: Classical Foundations for a Post-exemptionalist Sociology." *American Journal of Sociology*, 117(6): 1625–1673.

Frank, David John, Ann Hironaka and Evan Schofer. 2000. "The Nation-State and the natural environment over the twentieth century." *American Sociological Review*, 6(1): 96–116.

Franzen, Axel and Reto Mayer. 2010. "Environmental attitudes in cross-national perspective: A multilevel analysis of the ISSP 1993 and 2000." *European Sociological Review*, 26(2).

Galt, Ryan E., Leslie C. Gray, and Patrick Hurley. 2014. "Subversive and Interstitial Food Spaces: Transforming Selves, Societies, and Society–Environment Relations through Urban Agriculture and Foraging." *Local Environment*, 19(2): 133–146.

Galt, Ryan E., Leslie C. Gray, and Patrick Hurley. 2014. "Subversive and Interstitial Food Spaces: Transforming Selves, Societies, and Society–Environment Relations through Urban Agriculture and Foraging." *Local Environment*, 19(2): 133–146.

Goodman, Michael. 2015. "Food Geographies I: Relational Foodscapes and the Busy-ness of being More-than-food." *Progress in Human Geography*, 40(2).

Haluza-DeLay, Randolph. 2008. "A Theory of Practice for Social Movements: Environmentalism and Ecological Habitus." *Mobilization: An International Quarterly*, 13(2): 205–218.

Harvey, David. 2017. *Marx, Capital, and the Madness of Economic Reason*. New York: Oxford University Press.

Hayes-Conroy, Allison and Jessica Hayes-Conroy. 2008. "Taking Back Taste: Feminism, Food and Visceral Politics." *Gender, Place, and Culture*, 15(5): 461–473.

Hayes-Conroy, Jessica and Allison Hayes-Conroy. 2010. "Visceral Geographies: Mattering, Relating and Defying." *Geography Compass*, 4(9): 1273–1283.

Hayes-Conroy, Jessica, 2011. "School Gardens and 'Actually Existing' Neoliberalism." *Humboldt Journal of Social Relations*, 33(1/2): 64–96.

Holt, Douglas B. 1998. "Does Cultural Capital Structure American Consumption?" *Journal of Consumer Research*, 25(1): 1–25.

Inglehart, Ronald. 1995. "Public Support for Environmental Protection: Objective Problems and Subjective Values in 43 Societies" *Political Science and Politics*, 28(1): 57–72.

Isenhour, Cindy, Mari Martiskainen and Lucie Middlemiss. 2019. *Power and Politics in Sustainable Consumption Research and Practice*. Philadelphia: Routledge. [VitalSource Bookshelf]. Retrieved from https://bookshelf.vitalsource.com/#/books/9781351677301/.

Johnston, Josée and Shyon Baumann. 2015 [2010]. *Foodies: Democracy and Distinction in the Gourmet Foodscape* (2nd ed.). New York: Routledge.

Johnston, Josée and Shyon Baumann. 2007. "Democracy versus Distinction: A Study of Omnivorousness in Gourmet Food Writing." *American Journal of Sociology*, 113(1): 165–204.

Kasper, Debbie V. 2009. "Ecological Habitus: Toward a Better Understanding of Socioecological Relations." *Organization & Environment*, 22(3): 311–326.

Kennedy, E.H., J. Johnston and J. Parkins. 2017. "Small-p politics: How Pleasurable, Convivial, and Pragmatic Political Ideals Influence Engagement in Eat-Local Initiatives." *British Journal of Sociology*. doi:10.1111/1468-4446.12298.

Kennedy, Emily H., Shyon Baumann and Josee Johnston. Forthcoming. "Eating for Taste, Eating for Change: Comparing Cultural Capital in Foodie and Ethical Consumer Orientations."

Kurtz, H.E., 2001. "Differentiating Multiple Meanings of Garden and Community." *Urban Geography*, 22(7): 656–670.

Lamont, Michele. 1992. *Money, Morals, and Manners: The Culture of the French and the American Upper-Middle Class*. Chicago: University of Chicago Press.

Maguire, Jennifer Smith, ed. 2017. *Food Practices and Social Inequality: Looking at Food Practices and Taste across the Class Divide*. New York: Routledge.

Marx, Karl. 1990 [1867]. *Capital: Volume I*. Translated by Ben Fowkes. London: Penguin Books.

McClintock, Nathan. 2014. "Radical, Reformist, and Garden-Variety Neoliberal: Coming to Terms with Urban Agriculture's Contradictions." *Local Environment*, 19: 147–171.

McEachern, Morven G., Gary Warnaby, Marylyn Carrigan and Isabelle Szmigin. 2010. "Thinking Locally, Acting Locally? Conscious Consumers and Farmers' Markets." *Journal of Marketing Management*, 26(5–6): 395–412.

McMichael, Phillip. 2012. *Development and Social Change: A Global Perspective*, 5th ed. Sage: Los Angeles.

Megicks, Phil, Juliet Memery and Robert J. Angell. 2012. "Understanding Local Food Shopping: Unpacking the Ethical Dimension." *Journal of Marketing Management*, 28(3–4): 264–289.

Micheletti, Michele and Dietlind Stolle. 2005. "Swedish Political Consumers: Who They Are and Why They Use the Market as an Arena for Politics." pp. 145–164. In *Political Consumerism: Its Motivations, Power, and Conditions in the Nordic Countries and Elsewhere*, edited byMagnus Boström, Andreas Følles-dal, Mikael Klintman, Michele Micheletti, and Mads Sørensen. Proceedings from the 2nd International Seminar on Political Consumerism, Oslo August 26–29, 2004. Oslo: Nordic Council of Ministers.

Middlemiss, Lucie. 2018. *Sustainable Consumption*. Philadelphia: Routledge. [VitalSource Bookshelf]. Retrieved February 1, 2020 from https://bookshelf.vitalsource.com/#/books/9781317239819/.

Mincyte, D. and K. Dobernig. 2016. "Urban Farming in the North American Metropolis: Rethinking Work and Distance in Alternative Agro-Food Networks." *Environment and Planning A*, 48(9): 1767–1786.

Neilson, Lisa A. and Pamela Paxton. 2010. "Social Capital and Political Consumerism: A Multilevel Analysis." *Social Problems*, 57(1): 5–24.

Nevarez, Leonard, Kathleen Tobin and Eve Waltermauer. 2016. "Healthy Food Acquisition in a Food-Insecure City: An Examination of Socioeconomic and Food-Security Predictors." Presented at the ASFS/AFHVS conference, Burlington, Vermont, June 21, 2014.

Newman, Benjamin J. and Brandon L. Bartels. 2011. "Politics at the Checkout Line: Explaining Political Consumerism in the United States. *Political Research Quarterly*, 64(4): 803–817.

Nielsen, Annemetter and Lotte Holm. 2016. "Making the Most of Less: Food Budget Restraint in a Scandinavian Welfare Society." *Food Culture and Society: An International Journal of Multidsciplinary Research*, 19(1). doi:10.1080/15528014.2016.1145003.

Ocejo, Richard E. 2017. *Masters of Craft: Old Jobs in the New Urban Economy*. Princeton, NJ: Princeton University Press.

Polanyi, Karl. 1944. *The Great Transformation: The Political and Economic Origins of Our Time*. Boston, MA: Beacon Press.

Schoolman, Ethan D. 2017. "Building Community, Benefiting Neighbors: 'Buying Local' by People who do not fit the Mold for 'Ethical Consumers.'" *Journal of Consumer Culture*. doi:1469540517717776.

Teig, E., J. Amulya, L. Bardwell, et al. 2009. "Collective Efficacy in Denver, Colorado: Strengthening Neighborhoods and Health through Community Gardens." *Health & Place*, 15: 1115–1122.

Veen, E.J. 2015. "Community Gardens in Urban Areas: A Critical Reflection on the Extent to which they Strengthen Social Cohesion and Provide Alternative Food." PhD Thesis, Wageningen University, the Netherlands.

Weber, Max. 1930. *The Protestant Ethic and the Spirit of Capitalism*. Translated by Talcott Parsons. New York: Scribner.

West, Patrick C. 1985. "Max Weber's Human Ecology of Historical Societies." pp. 216–243. In *Theory of Liberty, Legitimacy and Power: New Directions in the Intellectual and Scientific Legacy of Max Weber*. Boston: Routledge & Kegan Paul.

Willis, Margaret M. and Juliet B. Schor. 2012. "Does Changing a Light Bulb Lead to Changing the World? Political Action and the Conscious Consumer." *The ANNALS of the American Academy of Political and Social Science*, 644(1): 160–190.

Wright, Erik Olin. 2010. "Interstitial Transformations." Chapter 10 in *Envisioning Real Utopias*. London: Verso.

5 "It connects me to the Earth"
Marginalized environmentalism and a resistance to capitalist logic

I met Marty[1] at his home in rural Illinois. I was welcomed by his wife Linda to their modest living room where they take seats in twin beige recliners that face a large television set under a wall of gold-framed family photos. They offered me some of their homemade wine. They were proud of it, and as I sipped it I made sure to compliment it profusely. Showing characteristically Midwestern humility, Marty described his wine as "just good enough to drink." Marty and Linda's pantry was lined with home-canned food from the garden; in the garage there were stacks of homemade wine, and the chest freezer was filled with game and fish. As Marty and I sat down to chat early on in my data collection process, my pre-formed schema came into play: he seemed like a rural, working-class conservative. I wrote in my field notes: *Marty is in Carhartts, pants and jacket.* One thing was clear: I did not conceptualize him as a typical environmentalist.

But as the conversation developed, my standpoint began to shift. On the topic of fishing Marty expressed his concern about pollution, species decline, fish lifecycles, nuclear radiation, among other deeply environmental topics. I saw that, despite his strong opposition to political environmentalism (which he, alongside environmental sociologists (Boucher 2017; Guber 2012; Leiserowitz et al., 2012; Nisbet and Myers, 2007), have typically associated with concerns over climate change), he is in fact embedded in and therefore cares deeply about the ecosystems that provide him not only food, but joy.

Since at least the beginning of the popular environmental movement, sociologists have focused on the ways in which humans have had major impacts on their environments as well as the responses that have arisen to address environmental problems. However, in their investigation of responses, researchers have long focused on self-proclaimed environmentalists to identify the ways in which people are navigating solutions to environmental problems (Franzen and Mayer 2010;

Inglehart 1995). Yet, political environmentalists represent a small subset of the world population that tend to be white, high SES, left-leaning and highly educated (Boli and Thomas 1997; Frank, Hironaka and Shofer 2000). Literature also shows that those who identify as environmentalists tend to be in the group of greatest greenhouse gas emitters when compared with the overall world average (Boucher 2017), as a function of their socioeconomic status. The current research demonstrates the researchers of sustainable consumption should cast their gaze beyond predominantly white, educated, wealthy environmentalists (Boli and Thomas 1997; Boucher 2017; Frank, Hironaka and Shofer 2000; Franzen and Mayer 2010; Inglehart 1995) toward populations that may be on the forefront of innovating pro-environmental solutions (Blake 1999; Hoggett 2013; Kollmuss and Agyeman 2002; Whitmarsh, Seyfang and O'Neill 2011).

The data of this study reveal the surprising ways in which both self-identified environmentalists and non-environmentalists alike care deeply about the hyper-local environment that produces their food due to their proximity to it (McClintock 2014; Mincyte and Doebernig 2016). I argue this is due to susbsistence food producers (SFPers) ecological embeddedness, or "the degree to which [an individual] is rooted in the land... [both being present] on the land and learns from the land in an experiential way" (Whiteman and Cooper 2000, 1267). The current research aims to add to the call for inquiry on alternatives arising as social crises unfold (Beck 2016; Berman 2017; Gibson-Graham 2006; Harvey 2017; Urry 2010). Building on previous chapters, the data suggest that in the development of subsistence food production as a neo-Polanyian shadow structure, individuals are developing pro-environmental practices by connecting to nature through the act of food production.

Ecological embeddedness

Although Marx's theories of ecology are not what he is most known for, throughout his work you will find a discussion of man's alienation from nature and the material conditions of existence (Foster 1999). Discussed in the Chapter 3, this alienation described in detail by Marx includes social and ecological estrangement (Galt, Gray and Hurley 2014; Marx 1990; McClintock 2014) as well as distance from manual work (Mincyte and Doebernig 2016). Marx's alienation is also related to Weber's disenchantment, arising from the changing relationships in society from Gemeinschaft to Gesellschaft (Berman 1981; Foster and Holleman 2012; Weber 1930). The data from the current study suggest that alienation from work and the material conditions of one's existence are countered through producing one's own food. I use ecological embeddedness theory, which is in many ways the direct counter to ecological estrangement (Foster 1999), to explain these findings.

The literature on ecological embeddedness in food systems has developed out of the interrogation of alternative food networks. These alternatives often seek to counter industrial agricultural practices, which inherently distance consumers from food production (Murdoch and Miele 1999; Murdoch, Marsden and Banks 2000; Nygard and Storstad 1998; Whatmore and Thorne 1997). In a refinement and development of the concept, Morris and Kirwan (2011), focusing mainly on farmers that produce food for sale, claim that ecological embeddedness is most helpful in explaining the ways in which food producers relate to and conceptualize the ecological processes on which they rely for food production. In order to be considered ecologically embedded, practices must be informed by ecology in some way and the values associated with the ecology must tend toward *conservation* and *sustainable use of the land* (Morris and Kirwan 2011).

Most food in the United States is produced through industrial agriculture, which has a standardized set of practices meant to bolster efficiency at the expense of conservation or sustainability. On the other hand, subsistence food producers reject standardization and instead produce food based on the specific, individual material constraints they face in their particular geography. The data suggest that subsistence producers are forced to make decisions about how to grow food, what inputs to use, and can no longer outsource those decisions by buying food at the supermarket grown by industrial producers who make those decisions for them. I argue that this reckoning with the difficult decisions of food production is due to ecological embeddedness (Galt et al. 2014; McClintock 2014; Mincyte and Dobernig 2016; Morris and Kirwan 2011; Whiteman and Cooper 2000), which simply forces proximity to material conditions and ecological processes.

Food producers in my sample report making choices that demonstrate a deep care for the environment that produces their food including composting, water use reduction, promoting biodiversity through diverse agroecology and refraining from chemical inputs, organic pest/disease management, and conservation-minded land management for hunters and fishermen. These practices contrast starkly from industrial food production practices that SFPers could choose to incorporate into their small-scale production. Instead, the data suggest that SFPers make more sustainable, conservation-minded choices due to their proximity to and their stake in the ecological systems that produce food, which is oftentimes their own property or a natural environment (lakes, oceans, forests) with which they have an ongoing relationship.

The concept of ecological embeddedness can then be operationalized to help to explain certain aspects of SFP. First, ecological embeddedness as a concept can help to explain the heterogeneity of alternative food network practices. If each ecological system that supports food production has its own set of dilemmas and structures, resultant practices will be situated in a diverse variety of contexts, in contrast to standardized industrial agricultural practices that use industrial processes to ignore ecological specificities to the extent possible. The concept can also be used to understand the ways in which ecological knowledge is produced through proximity to and ongoing use of a space (Morris and Kirwan 2011). In my results, I build on the concept of ecological embeddedness put forth by Morris and Kirwan (2011) and Whiteman and Cooper (2000) to show the ways in which SFPers negotiate and develop diverse site-specific practices through a care for and relationship with the ecological system that sustains food production. Overall, I add a more nuanced understanding of neo-Polanyian shadow structure development by demonstrating that subsistence food producers connect to nature and act to protect it through ecologically embedded food production.

Ecological embeddedness of environmentalists

Carter and Lydia are an upper-class couple who had taken to produce food for their own consumption by means of a large vegetable garden, a large flock of chickens, keeping fruit trees, bees, and fishing. They live in a wealthy suburb of Chicago with pristinely landscaped homes on large lots. As they let me into their home, they escorted me to a sitting room with overstuffed leather chairs and oil paintings on the walls. As the literature shows, upper SES people are more likely to hold environmental identities (Boucher 2017), and Carter and Lydia were very explicit about their environmentalist identities, as evidenced by Lydia:

> Who decided we should all have lawns? Who decided we can't have clover and dandelions on our lawns? Some commercial entity concluded that this was a good idea and then we keep drinking the Kool-Aid. It takes everything I have not to say to my neighbors 'get used to what we're doing to our lawn...With global warming

the algae is going to bloom and we're all going to be in trouble.'
Our lawn is like dandelion land surrounded by perfect green.

Here Lydia related her choice to refrain from chemical use at home
with larger-scale environmental issues such as agricultural runoff, algal
blooms, and climate change. This aligns with the literature that shows
belief in and care for issues of climate change is significantly associated
with higher SES (Dietz et al. 2007; Leiserowitz et al. 2012; McCright
and Dunlap 2011).

Through the process of producing their own food, Carter and Lydia
started to become aware of what kinds of SFP practices they wanted to
take part in. Carter described his first experience growing food for his
own consumption in a community garden:

> I started gardening in a community garden. And that was a real
> problem because my neighbors right next to me would spread
> herbicide and all that. I thought that was a real injustice. Like, it's
> not like I care what you do, but if it impedes on me I do care.

Carter explained his stake in and care for the environment that produces his food. Due to his proximity, or ecological embeddedness, he
became aware of and took responsibility for the practices that could
have impeded on his perceived land quality. Through this process
of ecological embedding, Carter developed an ethos or values-system
that promoted conservation and sustainable land use, according to
Morris and Kirwan's parameter (2011). This is evidenced through
his disdain for the use of herbicide and its impact on biodiversity
and ecological health, and his choice to use organic pest and disease
management.

Lydia describes her process of coming to know agricultural practices:

> I mean people, my friends, would be like 'do you have a rooster?'
> 'No' 'Then how do you have eggs?' People just don't know. I
> didn't even think about the whole cow thing. I was like 'ok the
> cow has to be pregnant before it makes milk, and then what
> happens to the babies?' I knew the cow made milk, but the closer
> we get to the earth in the process the more I see the disconnect
> [that other people have in their understanding].

As they took on more responsibility for the choices that led to the
production of their food, they not only became more aware of the

processes involved, but felt responsible for them. As was highlighted in the last chapter, Lydia explained that it was the sheer amount of experience involved in the process of self-producing that has impacted her sense of proximity to nature, "There's something very – it connects me to the earth. I go out every day. I look at the garden, I look at the sky, I look at the trees, I watch the cycle of the seasons happen because I am caring for these farm animals." This is an example of the ongoing relationship Carter and Lydia have developed with the land that supports their food production.

As the relationship became deeper and more complex, so did their ecological embeddedness, which developed into changed practices:

> I also think our neighbors, they spray their lawn and they put weed killer down. And we don't do any of that. It has opened up a dialogue with my neighbor. He says 'everything you plant grows. Nothing I plant ever grows'...What has happened in our garden that's been really miraculous to me is a gradual development of biodiversity. In the beginning, we had some bees. Now we have 12 or 14 different kinds of bees that show up in different parts of the season: single, solitary squash bees, and little minute black bees that come only in strawberry season, or moths. The diversity that we get that has exploded over time, has matured over time, gets more every year. Birds. Butterflies. That is something that they [my neighbors] miss. They get that there's something different happening here. I don't care if I have dandelions.

The data here suggest that this choice to self-produce leads to both an awareness of agricultural processes and a deep care for the material and ecological conditions that allow for food production, or ecological embeddedness (Galt et al. 2014; McClintock 2014; Mincyte and Dobernig 2016; Morris and Kirwan 2011; Whiteman and Cooper 2000). Carter and Lydia's choice not to spray chemicals on their lawn or garden is a practice that is not the social norm in their suburban community. Yet, they make this choice and see the direct result of that choice in their backyard as an explosion of biodiversity.

Jenny is an upper-class white urban woman with a large vegetable garden and chickens who has been marked as an environmentalist within the sample. Earlier in the interview, Jenny told me she felt disconnected from the process of food production. Upon getting chickens, she was made aware of the ecological and visceral processes involved in the production of eggs. She described a day when she went out with her young son, Elliot, to gather eggs:

When he goes out he's got this special little basket with handles that he can hold perfectly...he takes it very seriously, his job of like gathering the eggs...One time we actually stood there, Elliot and I, and we watched one of our hens lay an egg [laughter]...it seems like if we stand here we might get to see it...So we were standing there and all the sudden we see like *clunk* and it hit the base of the nesting box and we were both like [gasp]. We just kind of hugged and like cried a little bit, it was so emotional. It was really neat.

Here Jenny described this lived, very visceral (Hayes-Conroy 2011) experience of watching what will become her food being produced, and the dramatic emotional response. This physicality in the developing relationship with the ecosystem is something that has not been yet discussed in detail in the literature on ecological embeddedness. Yet it is present throughout the data: a sense of tangibility and experiential learning is central to the development of ecological embeddedness within this community.

I argue that this deep connection to nature afforded by the proximity of the production in SFP leads to a sort of virtuous cycle in which behavioral choices are informed by the connection to land, plants and animals. Yet, not every aspect of ecological embeddedness is a positive experience. As Jenny was new to keeping chickens, I asked her about what will happen when a chicken gets ill:

I do worry. It's hard. I've had childhood dogs get sick and have to be euthanized. It's going to be awful when it happens and all the stuff I read beforehand was like 'Be prepared. Chickens die and be prepared, be emotionally prepared.' It will be terrible when it happens but it'll be okay...So yeah I do think part of it's the circle [of life]. So we talk to Elliot about death, so he is prepared.

Jenny faced the reality of the choice of what to do with the death of the animal that produces her food, and it was difficult for her to emotionally process having to make the decision to euthanize an animal who cannot survive any longer. Yet, she took on the task of that responsibility in a way that attempts to bridge alienation from the material conditions of her existence (Foster 1999), or to enact a sort of re-enchantment, to counter Weber's concept of disenchantment (Berman 1981; Foster and Holleman 2012), connecting Jenny to the natural processes involved in food production.

Emma and Bill are a lower-class middle-aged white couple who live on a suburban lot and keep chickens, have fruit trees and bushes, have a large vegetable garden, forage, glean, hunt, fish, keep bees, and make beer and wine. They practice a range of pro-environmental behaviors including reducing overall consumption of consumer goods, recycling, composting, reducing water consumption, and organic pest and disease management. They described to me that the learning process of growing food is experimental – it is mainly a result of trial and error. Indeed, it is the experience of being in the garden, more so than any other source of knowledge, that has informed the practices they have developed. One aspect of gardening that all gardeners are faced with is water use. Emma described her process for watering:

> We have four rain barrels. We have drip hose and we put it on a timer. Last summer I was on a kick to reduce water consumption and we had a lot of rain so I only use the hose on the timer when we're not home. Everything is mulched so that the water is just going to the roots of the plants. So we get away with as little water as we possibly can. All the rest is hand watering from the rain barrels.

It was due to Emma's daily interaction with her garden that she was able to pay close attention to her water use and decide when to use the timer and when to use the rain barrel. Because she was in control of the process, she could bypass the timer when there had been a lot of rain, and use the full rain barrels for a period of time. She had implemented the use of mulch and drip tape – that keeps water from evaporating and goes directly to the roots of the plant where the water is taken up – thereby wasting very little water in the process.

Emma's process for dealing with pests and diseases in the garden was similar, "Well there's vinegar and there's pepper and there's hand picking them. There's rotating crops, there's proper mulching, careful watering. If your plants are healthy you're less likely to have pests." Over 23 years of gardening, Emma had learned how to manage gardening problems in a way that does not use standardized chemical inputs in the manner of industrial agriculture. Even more profoundly, through ecological embeddedness, she had learned how to notice the micro-variations in plants, rainfall, soil quality, etc. That knowledge allowed her to keep her garden healthy through appropriate watering and maintaining healthy soils, thereby negating the need for chemical inputs.

It is the act of subsistence food *production*, which is fundamentally different from sustainable *consumption*, that forces individuals to make decisions about how to hunt, garden, keep chickens, handle waste, or fish. Producers are confronted with the realities of how food is made and are forced to make choices about how to do it. These producers are not industrial farmers, and have no training in standardized agricultural practices. Instead, SFPers have to learn through trial and error, or by interacting with their land through a form of ecological embeddedness. As this relationship with the ecosystem develops, a sort of visceral re-enchantment happens that not only results in a greater awareness of natural processes, but also the consideration of ecological outcomes (Berman 1981; Morris and Kirwan 2011; Whiteman and Cooper 2000). The end result is self-reported pro-environmental behaviors such as landfill diversion/ closed-loop waste processing, organic pest/disease management, lowered overall consumption of industrial foods, seed saving, soil remediation, recycling, and more.

In many ways, it is expected that those holding environmentalist identities would act to produce food in ways that protect the environment and tend toward principles of conservation. In this section, I explored the way in which the act of production forces environmentalists to take responsibility and make decisions about their food production, thereby leading to changed behaviors and sustainable practices. What is surprising, however, is that this same kind of logic and set of practices also apply to non-environmentalists in my sample. In the next section, I will explore the unexpected way in which those individuals who are often left out of the discussion of environmental solutions – who tend to be lower-class, rural, conservative, or people of color – also are ecologically embedded in their subsistence food production practices, and the implications this has for future research and activism.

Ecological embeddedness of non-environmentalists

Marty, mentioned in the introduction, is a white rural working-class commercial flower farmer in his late fifties who provides most of his food consumption, especially in the summer months, through a combination of fishing, hunting, vegetable gardening and fruit trees, as well as producing his own wine. When I spoke to Marty, he explained that he was wary of environmentalists, and disdainful of the idea of climate change:

> The oceans aren't that much warmer than they ever were. You read about the falsified data. They get these grants and if you come back and say 'oh, it's not happening' then no more money!

We need more money to further this and then the political agenda is fulfilled and it's too bad.

Literature shows that Marty is in the majority of Americans in holding this belief, as only a minority believe climate change is both real *and* caused by human activity (Boucher 2017; Guber 2012; Leiserowitz et al. 2012; Nisbet and Myers 2007).

However, Marty is an avid fisherman, and cares deeply about protecting fish populations and marine environments. Marty described his experience fishing in the Florida Keys:

> The cruise ships are the biggest thing there. They just dump [waste]. We fish the Florida Keys. We've watched brown water come in, and human waste come in, and plastic bags that are all ground up come in and they say they'll come back and clean it all up but that's a lie. We've seen gigantic barges pulling out mountains of garbage and they get out and they just dump it.

Marty espoused values that were clearly aligned with those of ecological embeddedness such as conservation and sustainability, lamenting the pollution issues that affect this ecosystem he cared deeply about. This suggests that the inclusion criteria of belief in human caused climate change for "environmentalist," may be leaving out populations who have deep care for other environmental issues such as conservation and pollution efforts.

Not only did Marty care about the ecosystem, he also cared deeply about protecting fish populations:

> The commercial [fishermen] are exempt from these limits on taking out fish. The commercial guys have more political clout because their licenses are such. If you go out on a boat fishing and you are only allowed five [fish] a piece and the captain is allowed to get sixty and he goes back and sells them to the fish market, well how is that good? They are trying to stop that, but there's too much money involved. Once money gets involved, you're not going to shut it down. The oceans are overfished, there's no question about it.

In this excerpt, we see Marty making a connection between capitalist gain and the power associated with access to these fishing resources, at the expense of fish populations. In lamenting the practices of other fishermen, Marty is distancing himself from industrial or larger-scale methods which he sees as fundamentally harmful to this ecosystem on which he relies for food. This is an illustration of a sense of disenchantment from and skepticism for processes of hegemonic capitalist ways of producing food, and a tendency toward practices that take into account the health of the natural world (Berman 1981).

Later, Marty got into more detail not only about fish populations but about the details of fish life cycles:

> As species start declining – well you can already see it. For instance, the wahoo that we shoot, 50 pounds is a big one. 70 pounds is a real big one. They used to average 140 pounds. There are no 140 pounders anymore. They are taking them so fast that they are not allowed to get to full maturity. The swordfish are the same way. The swordfish used to be a couple thousand pounds. Now the swordfish are on average 100–150 pounds. The blue fin tuna, they used to go out and catch boatloads of them, not anymore. And the only reason there are a couple big ones left is because no one can get to them, you need to have the right equipment, and commercial guys can do that. I think one of the few places in the world where the Bluefin spawn is off the coast of Japan, now with this radiation from Fukushima is coming off the coast. It's a mess...And it's not climate change.

This quote highlights some important themes. First, Marty clearly identifies himself as in opposition to environmentalists. He sees elite environmental issues as influenced by power and corruption in politics. Instead, Marty practices a sort of environmental concern for the immediate ecosystem that impacts his food production. He displays a deep understanding of this marine ecosystem, being concerned with pollution, radiation, overfishing, and even fish life cycles. But he does not forget to end his statement with "and it's not climate change," to cement his identity as differentiated from environmentalists.

Marty's concern for the marine environment is closely linked with his desire to access foods for his own consumption. When alienated from food sources (Berry 1977; Marx 1974; 1981; 1990), consumers have less awareness of or care for the practices involved in producing their food. However, as individuals take on the work of self-producing, they are then exposed directly to the specific practices involved. In the words of Kloppenburg, Hendrickson and Stevenson, "It is through food that humanity's most intimate and essential connections to the earth and to other creatures are expressed and consummated" (1996, 37). In Marty's case, he is concerned about the industrial practices that can negatively impact the quality of the marine environment in which the fish live, because those fish will eventually be put into his own body. This is evidence of his awareness of the integrated relationship between the material conditions of his existence (or the biophysical environment) and his own body (Ollman 1976; McClintock 2010). In addition, Marty also has a lived, visceral connection to the ecosystem he fishes. He speaks fondly of his fishing sites, and the moments of bliss he has experienced there.

This concern for or attention to micro-environments in the localized production of food was repeated by nearly all my participants in one way or another. Noah and Joann, a white working-class suburban couple with three small kids, keep ducks, vegetables, fruit trees, and barter for meat. They were classified as non-environmentalists in the sample. Yet, in discussing the process of assessing the quality of their backyard soil, they display a deep concern for pollution and toxicity in their soil:

Noah: So a good example is like lead content in your soil, right? 'Cause there's disagreement about whether in soil it's good or not good or how you find out and so there's contradictory information...our lot has been residential since day one and we're in the middle of a residential neighborhood so there's no industrial – no dry cleaners, there's no foundries. Like upstream from us or upwind from us. It's always been residential, our side-lots never been built on...Our black topsoil is a foot deep, it's amazing. But I'm not too worried about it. What's in it? It's exhaust from cars, it's paint chips from the garage, right? That's the kind of lead contamination. We sent it off and got tested and ours came in at 298 ppm, but I intentionally sampled areas next to the garage, next to the house where I knew paint chips had been falling for years. Am I worried about 298 parts? No, I'm not. At some point, you have to go with it and say –

Joann: Well at some point you have to be realistic about 'okay is the farmer that's growing the industrial food any more vigilant about contamination than we are?' No. [laughter]

Noah: Or coming from China?

This level of practical closeness to the food production environment means that participants are directly aware of the effect of toxins on their own produce.

Certainly, Noah and Joann sensed that distance from their food source meant a lack of information about the quality of the practices being enacted there. However, they were willing to accept "good enough" lead content in their own soil because it is a known quantity. When they undertake the work of home food production in their backyard, Noah and Joann are grounded in the realities of their soil, and having to embrace bodily the known consequences of toxicity helps to counteract the distancing of the industrial agro-food system as well as the abstraction that brings (Mincyte and Dobernig 2016). Their concern for and proximity to their food production has led them to choose organic methods of food production, refraining from chemical use, in order to maintain soil quality.

Danny, a young working-class urban man who produces food through a combination of fishing, hunting and vegetable gardening, describes his keen awareness of nature through the practice of hunting. Like Marty, Danny has a relationship with this specific ecosystem and the immediacy of his connection with his hunting experiences has brought his attention to the details of that environment, "The wind changes, different weather, raining, sunny, cloudy. If it's cold out are [the deer] going to be bedded down? Are they going to be eating? Is there a lot of food? Is there little food?" With this visceral connection to the environment, Danny is attentive to the weather, the season, the state of the forest, all which impact the behavior of the animals on whom he depends for food. This is evidence of a sense of not only proximity to this place, but attention to and connection with the particularities of that space through a visceral experience of it. I argue that this bond leads to a greater care for its conservation, detailed on p. 69.

Danny is also mindful of the population numbers for hunted animals: deer, elk, wild turkeys, and pheasants. He describes to me how the populations of each of these species is either abundant (e.g. deer) or slowly recovering (e.g. turkeys). He is cognizant of and takes very seriously the limits placed by the Illinois Department of Natural Resources. Danny has a care and respect for the animals he hunts. He describes wild turkeys as "very smart" and having "very good eyesight." He also takes care to hunt humanely. He describes hunting with a crossbow, "A really good shot with a bow would consider 50 yards a maximum shot. Because it does lose velocity after a while. You don't want to injure or maim any animal." The data suggest that this visceral and intimate relationship Danny has developed with his hunting ecosystem has led him to make choices about how to hunt that differ from hunting practices that are not conservation-minded.

Despite not identifying as an environmentalist, Danny undoubtedly behaves in a manner in keeping with conservation philosophies. The fact that he displays a feeling of connection to the ecosystem, expresses respect for the life therein, and acts to protect and preserve this natural space underscores this ecologically embedded value system despite a lack of environmentalist identification. In the framework of Whiteman and Cooper's (2000) work on ecological embeddedness of land management practices of the Cree Native Americans, Danny's quotes illustrate several important dimensions including: ecological reciprocity, ecological respect, ecological caretaking, and ecological experience (adapted from Table 1, p. 1275). In other words, Danny's description of his time spent hunting reveals a connection to and care for the ecosystem that provides for him. The most important implication of this, according to Whiteman and Cooper (2000), is sustainable land management that respects ecosystems and animal populations.

Similarly to those who explicitly identify as environmentalists, non-environmentalists are also expressing a deep care for the proximate natural systems on which they rely for food production, due to their ecological embeddedness (Morris and Kirwan 2011; Whiteman and Cooper 2000), and this connection leads to choices that differ greatly from the ways in which food is produced industrially. The non-environmentalists in my sample are reporting behaviors such as limiting game harvests to promote healthy animal populations, following humane hunting practices, and building soil quality through organic practices. The data suggest that it is a perceived visceral connection to and care for the ecosystem of small-scale production that leads them to make different, pro-environmental choices for food production.

This finding has significant implications for the further study of solutions to environmental problems. Environmental scholars and activists must acknowledge the potential of populations marginalized by politics and academia. The data in this section suggest that those interested in the development of a sustainable future may want to reimagine what constitutes an environmentalist by including those whose values are being impacted by ecological embeddedness.

Inclusive environmentalism

Being close to the source of their food by self-producing, subsistence food producers are directly confronted with the natural environment from which they produce food (McClintock 2014; Mincyte and Dobernig 2016), leading to values such as conservation and sustaining healthy ecosystems. This finding is significant because environmental researchers and activists have typically focused on those that identify as environmentalists, who tend to be white, wealthy, well-educated, and left-leaning (Boli and Thomas 1997; Frank, Hironaka and Shofer 2000) as the main source of alternatives or solutions. The data suggest that environmentalist identity is not a necessary condition to valuing practices that promote conservation and sustainability. Some of the reported food production practices include: composting, water saving/reduction, food localization, closed-loop nutrient cycling, soil remediation, hunting and fishing conservation, and promoting biological diversity.

Those seeking to study human–environment interaction should look beyond those that simply identify as environmentalists for ecological innovation. In other words, it is the hunters and fishermen, backyard gardeners, and urban chicken keepers who have been marginalized in mainstream environmental sociological research who may be at the forefront of pro-environmental innovation, and the implication is that we must begin to look at these groups in order to learn about exciting directions in environmental research. Like Hochschild argues in her recent book *Strangers in their Own Land* (2016), we must take an empathetic look at the structural conditions of late capitalism that have alienated and disenfranchised diverse populations (Foster 1999), from urban communities of color to rural conservatives. We must then consider the ways in which these diverse communities are finding resilient, creative and innovative solutions that are, surprisingly, similar to one another. If we can move beyond those meanings – like environmentalist identity – that divide these populations politically, we could potentially find ways to work together practically to envision a more sustainable future.

Certainly, this finding is not assessing the ecologically beneficial outcomes of these processes and meanings, although small-scale food production has already been shown to have significant environmental and social benefits (Edmondson et al. 2014; Guitart, Pickering and Byrne 2012), and positive health outcomes, especially for low-income communities (Freeman et al. 2012; Kremer and DeLiberty 2011; Metcalf and Widener 2011; Wakefield et al. 2007). Subsistence food production has also been shown to produce healthier soils, when compared to industrial agriculture (Edmondson et al. 2014); as well as to provide substantial amounts of quality food for gardeners (Vitiello and Wolf-Powers 2014). To add to these findings, I suggest further interdisciplinary inquiry into the diverse set of practices described here by soil scientists, ecologists, medical researchers, and others.

What is clear from this chapter is that due to their proximity to the production, the food producers in my sample do indeed engage in concern for environmental best practices. Regardless of affiliation with environmentalism in the public or political sense, my participants have a deep care for their highly local environments due to their perceptions that the practices used in these places have a direct impact on their lives and health. This chapter adds to a more detailed understanding of shadow structure alternatives arising as social crises unfold (Beck 2016; Berman 2017; Gibson-Graham 2006; Harvey 2017; Urry 2010). In the exploration of alternative food production, subsistence food producers are both connecting to nature and developing best practices through an ecologically embedded lens. The findings of the current chapter suggest that self-producing food leads to the adoption of "rationalities that diverge from capitalist rationalities" (Galt et al. 2014, 137). In the process of production, those SFPers in my sample are becoming more acquainted with the ecological conditions of their existence, thereby countering capitalist logic in which distancing is inherent in the industrial food production system.

Note

1 All names are pseudonyms. All identifying details have been changed to protect anonymity.

Works cited

Beck, Ulrich. 2016. *The Metamorphosis of the World: How Climate Change is Transforming our Concept of the World*. New York: Polity Press.

Berman, Morris. 1981. *The Reenchantment of the World*. Ithaca, NY: Cornell University Press.

Berman, Morris. 2017. "Dual Process: The Only Game in Town." In *Are We There Yet?* Brattleboro, VT: Echo Point Books. Essay #27.

Berry, Wendell. 1977. *The Unsettling of America: Culture and Agriculture*. San Francisco: Sierra Club Books.

Blake, J. 1999. "Overcoming the Value-Action Gap in Environmental Policy: Tensions between National Policy and Local Experience." *Local Environment*, 4(3): 257–278.

Boli, John and George M. Thomas. 1997. "World Culture in the World Polity: A Century of International Non-Governmental Organization." *American Sociological Review*, 62(2): 171–190.

Boucher, J.L. 2017. "Culture, Carbon and Climate Change: A Class Analysis of Climate Change Belief, Lifestyle Lock-in, and Personal Carbon Footprint." *Socijalna Ekologija*, 25(1): 53–80.

Dietz, Thomas and Eugene A. Rosa. 1994. "Rethinking the Environmental Impacts of Population, Affluence and Technology." *Human Ecology Review*, 1: 277–300.

Dietz, Thomas, Eugene A. Rosa and Richard York. 2007. "Driving the Human Ecological Footprint." *Frontiers in Ecological and Environmental Science*, 5(1): 13–18.

Edmondson, Jill L., Zoe G. Davies, Kevin J. Gaston, and Jonathan R. Leake. 2014. "Urban Cultivation in Allotments Maintains Soil Qualities Adversely Affected by Conventional Agriculture." *Journal of Applied Ecology*, 1–10.

Foster, John Bellamy. 1999. "Marx's Theory of Metabolic Rift: Classical Foundations for Environmental Sociology." *The American Journal of Sociology*, 105(2): 366–405.

Foster, John Bellamy and Hannah Holleman. 2012. "Weber and the Environment: Classical Foundations for a Post-exemptionalist Sociology." *American Journal of Sociology*, 117(6): 1625–1673.

Frank, David John, Ann Hironaka and Evan Schofer. 2000. "The Nation-State and the Natural Environment over the Twentieth Century." *American Sociological Review*, 6(1): 96–116.

Franzen, Axel and Reto Mayer. 2010. "Environmental Attitudes in Cross-national Perspective: A Multilevel Analysis of the ISSP 1993 and 2000." *European Sociological Review*, 26(2).

Freeman, C. et al. 2012. "My garden is an expression of me": Exploring Householders' Relationships with their Gardens." *Journal of Environmental Psychology*, 32(2): 135–143.

Galt, Ryan E., Leslie C. Gray, and Patrick Hurley. 2014. "Subversive and Interstitial Food Spaces: Transforming Selves, Societies, and Society–Environment Relations through Urban Agriculture and Foraging." *Local Environment*, 19(2): 133–146.

Gibson-Graham, J.K. 2006. *A Postcapitalist Politics*. Minneapolis: University of Minnesota Press.

Guber, Deborah Lynn. 2012. "A Cooling for Climate Change? Party Polarizations and the Politics of Global Warming." *American Behavioral Scientist*, 57(1): 93–115.

Guitart, D., C. Pickering and J. Byrne. 2012. "Past Results and Future Directions in Urban Community Gardens Research." *Urban Forestry & Urban Greening*, 11: 364–373.

Harvey, David. 2017. *Marx, Capital, and the Madness of Economic Reason*. New York: Oxford University Press.

Hayes-Conroy, Jessica, 2011. "School Gardens and 'Actually Existing' Neoliberalism." *Humboldt Journal of Social Relations*, 33(1/2): 64–96.

Hochschild, Arlie Russell. 2016. *Strangers in their Own Land*. New York: The New Press.

Hoggett, P. 2013. "Climate Change in a Perverse Culture." pp. 56–71. In S. Weintrobe (ed.), *Engaging with Climate Change: Psychoanalytic and Interdisciplinary Perspectives*. New York: Routledge.

Inglehart, Ronald. 1995. "Public Support for Environmental Protection: Objective Problems and Subjective Values in 43 Societies." *Political Science and Politics*, 28(1): 57–72.

Kloppenburg, Jack Jr., John Hendrickson and G.W. Stevenson. 1996. "Coming in to the Foodshed." *Agriculture and Human Values*, 13(3): 33–42.

Kollmuss, A. and J. Agyeman. 2002. "Mind the Gap: Why Do People Act Environmentally and What are the Barriers to Pro-Environmental Behavior?" *Environmental Education Research*, 8(3): 239–260.

Kremer, P. and T.L. DeLiberty. 2011. "Local Food Practices and Growing Potential: Mapping the Case of Philadelphia." *Applied Geography*, 31(4): 1252–1261.

Leiserowitz, Anthony, Edward Maibach, Connie Roser-Renouf and Jay D. Hmielowski. 2012. "Extreme Weather, Climate and Preparedness in the American Mind." *Yale Project on Climate Change Communication and George Mason University Center for Climate Change Communication Report*.

Marx, Karl. 1974. *Early Writings*. New York: Vintage.

Marx, Karl. 1981 [1863–1865]. *Capital: Volume III*. New York: Vintage.

Marx, Karl. 1990 [1867]. *Capital: Volume I*. Trans. Ben Fowkes. London: Penguin Books.

McClintock, Nathan. 2010. "Why Farm the City? Theorizing Urban Agriculture through a Lens of Metabolic Rift." *Cambridge Journal of Regions, Economy and Society*, 3: 191–207.

McClintock, Nathan. 2014. "Radical, Reformist, and Garden-Variety Neoliberal: Coming to Terms with Urban Agriculture's Contradictions." *Local Environment*, 19: 147–171.

McCright, Aaron and Riley Dunlap. 2011. "The Politicization of Climate Change and the Polarization in the American Public's Views of Global Warming, 2001–2010." *Sociological Quarterly*, 52(2): 155–194.

Metcalf, S.S. and M.J. Widener. 2011. "Growing Buffalo's Capacity for Local Food: A Systems Framework for Sustainable Agriculture." *Applied Geography*, 31(4): 1242–1251.

Mincyte, D. and K. Dobernig. 2016. "Urban Farming in the North American Metropolis: Rethinking Work and Distance in Alternative Agro-Food Networks." *Environment and Planning A*, 48(9): 1767–1786.

Morris, Carol, and James Kirwan. 2011. "Ecological Embeddedness: An Interrogation and Refinement of the Concept within the Context of Alternative Food Networks in the UK." *Journal of Rural Studies*, 27: 322–330.

Murdoch, J., T. Marsden and J. Banks. 2000. "Quality, Nature, and Embeddedness: Some Theoretical Considerations in the Context of the Food Sector." *Economic Geography*, 76: 107e125.

Murdoch, J. and M. Miele. 1999. "'Back to Nature': Changing 'Worlds of Production' in the Food Sector." *Sociologia Ruralis* 39: 465e483.

Nisbet, M.C. and T. Myers. 2007. "The Polls: Trends. Twenty Years of Public Opinion about Global Warming." *Public Opinion Quarterly*, 71: 444–470.

Nygard, B. and O. Storstad. 1998. "De-globalisation of food Markets? Consumer Perceptions of Safe Food: The Case Study of Norway." *Sociologia Ruralis*, 38: 35–53.

Ollman, Bertell. 1976. *Alienation*. London: Oxford University Press.

Urry, John. 2010. "Consuming the Planet to Excess." *Theory, Culture and Society*, 27 (2–30): 191–212.

Vitiello, Dominic and Laura Wolf-Powers. 2014. "Growing Food to Grow Cities? The Potential of Agriculture for Economic and Community Development in the United States." *Community Development Journal*, 1–16.

Wakefield, S. et al. 2007. "Growing Urban Health: Community Gardening in South-East Toronto." *Health Promotion International*, 22(2): 92–101.

Weber, Max. 1930. *The Protestant Ethic and the Spirit of Capitalism*. Translated by Talcott Parsons. New York: Scribner.

Whatmore, S. and L. Thorne. 1997. *Nourishing Networks: Alternative Geographies of Food*. pp. 287–304. In D. Goodman and M. Watts (eds.), *Globalising Food: Agrarian Questions and Global Restructuring*. London: Routledge.

Whiteman, Gail and William H. Cooper. 2000. "Ecological Embeddedness." *Academy of Management Journal*, 43(6): 1265–1282.

Whitmarsh, L., G. Seyfang and S. O'Neill. 2011. "Public Engagement with Carbon and Climate Change: To What Extent is the Public 'Carbon Capable'?" *Global Environmental Change*, 21(1): 56–65.

6 "Without the garden we never would have met him"

Practitioner networks as post-capitalist shadow structures

Building on the findings of the last chapter that show subsistence food producers are informed by their proximity to nature to enact positive environmental behaviors, this chapter looks more deeply into the way information is shared horizontally across diverse social networks and the implications this has for social change. Just as in previous chapters, the data can be paradoxical. The embedded meanings associated with coming to subsistence food production (SFP) follow a neoliberal logic to enact private, individualized solutions to public issues, which is in line with previous scholarship of sustainable consumption (Johnston 2008; Szasz 2007).

Yet, as a result of seeking out help for problems that arise from the practice of SFP, people report developing a diverse network of practitioners, thereby moving from private solutions to more public ones. This finding diverges significantly from previous work on alternative food networks that look only at homogenous communities of practice (Block, Chavez and Allen 2011; Gottlieb and Joshi 2010; Mares and Pena 2011; Morales 2011). This book explores the way subsistence producers build social ties through practical, meaningful problem solving that in some cases leads to the development of political and economic shadow structures.

Subsistence food production as a vehicle of social change

It was mid-winter just after dark and I drove up to a coffee shop in a notoriously dangerous neighborhood on Chicago's far South Side to meet Madeline, a working-class black woman in her early seventies, who grows food in and runs several urban community gardens, hunts and fishes, and keeps animals including chickens, bees, ducks, goats, worms for making compost and a turkey in an exchange with a rural Illinois farm. I was nervous meeting Madeline at night. This neighborhood is inhabited almost exclusively by people of color, predominantly black, and I knew my presence was conspicuous as a white woman walking alone to the diner where we had agreed to meet. As I sat down with Madeline and we began to chat, I realized she grows food not only because she thinks it solves some of the practical issues she faces in accessing quality food in a food desert, but also because she sees it explicitly as a vehicle for social change.

She explained to me exactly how she sees the potential of SFP for her community. One year she worked with a community garden to produce enough food to sell to other households:

> We found some plans on Etsy for herb bottles. So we made thousands of them and gave them away to seniors with herb plants inside. We said 'stop buying Mrs. Dash! You have Mrs. Dash right here.' That's the kind of thing we want. To make change. If you have kids talking about change it will change them. Let me repeat: *if you have kids talking about change it will change them.*

Madeline believes the connection to work, nature, and community that results from subsistence food production can provide opportunities to members of her community. Madeline is in the minority of my sample in her explicit understanding of SFP as an act of intentional social change, but she has put her finger on something most of my respondents were doing without explicitly recognizing it.

Most participants began producing food as an individual solution to perceived problems that lie squarely in the public sphere – like access to quality foods. Yet, my results indicate that as interest and participation in food production develop, SFPers cultivate a network of practitioners with whom they share information and resources, which has several important implications – including the strengthening of civil society (Granovetter 1983; King 2008; Putnam 2000) and the development of shadow structures (Berman 2017) through interstitial strategies (Wright 2010), forming a kind of Polanyian double movement (1944).

Taking part in SFP follows a general timeline among my sample, as follows:

- Subsistence food producers start by getting exposure to SFP activity
- Have a gestation period of thinking about taking part in some SFP
- Make the decision to start SFP

- Encounter issues, and reach out for resources for help in problem solving parts of SFP
- In solving issues, develop a network of practitioners for troubleshooting, and cultivate ongoing relationships with other SFPers
- When needed, form public groups or mobilization of resources for social change.

For the sake of clarity, I will collapse the process into three parts, in temporal order: 1) perception of social problems and coming to SFP; 2) problem solving and developing social ties, and; 3) shadow structure development.

1. Perception of social problems and coming to SFP

Lucia, mentioned in previous chapters, is a working-class Latina living in an urban area who keeps a large vegetable garden, gleans, forages, barters and trades, and cultivates mushrooms. Describing why she began producing food, she explained that it was part of an outgrowth of her and her friends' disappointment with participatory politics. Here she described a conversation with her friend Mike where he contemplated what to best do with his passion to address issues of social inequality, environmental degradation, and the exploitation of disadvantaged people:

> [Mike] was in this place where he was kind of disillusioned...He really wanted to use the ideas that he had like built over time. And so we were just kind of like hanging out one night. And he realizes all of these amazing things [like community gardens] exist out here, none of that stuff's at home [on the South Side of Chicago]. I wanted to be at home where we have nothing. I mean, we need it here [in Chicago] more than Oakland [California] does. And that we have a personal connection to the community, and where we can relate to, and people talk like us. All of those wonderful things about being a Chicagoan.

Lucia explains she was being exposed to ideas through political participation in several organizations including Occupy Wall Street, but ultimately felt like she had reached a point at which pursuing her goals through more formal political channels had reached a dead end.

She and a group of friends decided that focusing their attention on subsistence food production was a specific, localized action that helped them to realize a solution to some of the perceived problems they got into politics to help solve. To them, growing their own food was a practical way to get started, to realize their vision, and they felt it was important to do this activity in a place they feel is under-represented in alternative food networks (unlike Oakland, California). Lucia and her small group of friends then decided to take over an abandoned lot (with permission from her City Councilperson) to start growing food.

George is a working-class black man living in a rural area, who produces his own food through a combination of vegetable gardening, keeping fruit trees and bushes, keeping chickens and goats, fishing, foraging for mushrooms and berries. He also makes his own biodiesel fuel. His process of coming to SFP was similar to Lucia's. Here he described coming to the realization that in the coming environmental and social crises, marginalized communities will be the worst affected, as has been demonstrated by sociology of disaster literature (Brunsma, Diverfelt and Picou 2007; Fussell 2008):

> I went to the conference [on peak oil] and it wasn't a race thing but they said who wasn't going to likely make it [out of impending environmental crises]. And guess who that was? That was the [low-income, black] community that I'm from! So I was like 'oh my god! What are we going to do about it?' You gotta let people know! You gotta let them know about energy and food so we can be self-sustaining so we can be prepared for this great change that's going to come that we can't stop. From that point, we came back home and started our sustainability center.

George is describing something that repeated consistently in the data: some sense risk from various impending crises (Beck 1992; 2016), and a sense of efficacy, safety or self-reliance by taking part in SFP. George describes how it was learning about peak oil and environmental injustices that made him decide to grow food and produce biodiesel as a way to be more "self-sustaining."

In some ways, this very closely mirrors the conclusions of several scholars of sustainable consumption in which individuals change their consumption patterns toward more "sustainable" consumption, and thereby feel protected from risk, channeling them away from more political activity into this neoliberal, individualized solution (e.g., Kennedy and Johnston 2014; MacKendrick and Stevens 2016). This was explored most famously in Andrew Szasz's book *Shopping Our Way to Safety* (2007) in which people enact individualized solutions (e.g., buying bottled water and organic vegetables) to solve perceived social problems (e.g., water and food contamination) as a way to feel in control. However in Szasz's work, like others who look at individualized sustainable consumption, the solution is green consumer action – choosing to *buy* products that are perceived to be safer and more environmentally friendly.

Szasz found that once individuals felt they were buying the right products, they could assuage their environmental concerns and therefore not take part in political action, which he calls *inverted quarantine*. Unlike the public health model of quarantine in which individuals are kept away from others to spread disease, inverted quarantine suggests that individuals choose to wall themselves off from risk through the perceived efficacy of consumer choice. In the case of subsistence food production, however, the changed behavior is significantly different from consumer action. Instead, the act of *production* (as we will see) forces people into the public sphere, through the necessary sharing of information and resources. Further, consumer activism is something that is accessible to the already privileged (Szasz 2007), and SFP has been demonstrated to be within reach for those less-privileged communities that most lack access to healthy foods (Block et al. 2011; Gottlieb and Joshi 2010; Morales 2011).

The decision to self-produce food can be explained as a Polanyian double movement (1944). Although I am limited to retrospective accounts of the decision to take part in SFP, the data suggest that my participants perceive not only a sense of risk, but also that the formal or political channels for addressing this sense of precarity are not effective. I argue this is a result of their perception of the extreme commodification of land, labor and money that Polanyi argued will push capitalism into crisis, and the result is a pushback in the form of double movement activity.

Madeline, the black woman mentioned at the beginning of this chapter, echoes the comments of Lucia and George. For her, as for most in my sample, SFP is an action that can offer pragmatic solutions (Kennedy, Johnston and Parkins 2017; McAdam et al. 2001) to the social problems she faces:

> We need to know that you can grow a lot of stuff in your own back yard that is more healthy than the stuff you get at the store. And there's the problem of inaccessibility. Everybody talks about deserts and stuff, we are trying to counteract that. There's stores, the filling stations, junk food, it's a whole commercial project that's been fed to us and we've absorbed it. And we've got to learn to spit that out.

In some important ways, the initial decision to take part in SFP represents a neoliberal, individual solution to perceived public issues, and confirms the literature on this subject (Johnston 2008; Szasz 2007). However, as the process of taking part in food production develops, it proves to be substantively different than shifting consumer choices, as it forces individuals to reach out to others to solve the problems that arise as a part of SFP.

2. Problem solving across dimensions of difference and developing ties

As SFPers begin to enact their specific production methods, they are inevitably faced with problems, even among those for whom SFP is something they have done their entire lives. As these problems arise, subsistence producers report reaching out to different resources such as books and websites. However, these resources are often found lacking, and this is when participants report frequently turning to communities of practice to share resources.

Discussed in the introduction, a group called Advocates for Urban Agriculture puts on a free exposition called the Urban Agriculture Livestock Expo each February. Mid-winter in a public high school on Chicago's south side there is an assemblage of people[1] as well as demonstrations and classes on animal-based food production including: chickens, bees, goats, guinea fowl, rabbits, ducks and turkeys. Despite the vast diversity in terms of social difference, people were able to immediately connect over problems of practice. I observed conversations with content related to: feeding and housing livestock, disease and pest remedies, concerns with weather, resources for quality products, sourcing of products, disposing of waste, processing/butchering, and breed selection. Most of the content of these practice-based conversations tended toward more organic or environmentally positive practices.

One interaction I noted was between a middle-aged white woman coming from a predominantly white working-class suburb[2] who keeps chickens, and a young, black, lower-class woman coming from a predominantly black neighborhood in the inner city who grows vegetables and is considering keeping chickens:

> The white woman pulls out her phone and shows pictures: This one is Margaret. She's a Rhode Island Red. Isn't she just beautiful?
> BLACK WOMAN: I had no idea how pretty these chickens could be.

WW: Oh yes, my girls are beauties. There's Rhonda, and Katie. [flipping through photos on her phone]
BW: So, how did you pick the breed?
WW: Well, the first thing you have the think about is the weather. You can't get no southern chickens up here; they won't make it through January. So, I went to this feed store over on Pulaski out south. I chatted with them until they were sick of seeing me, and I ordered the chicks through them.

(Field notes 14 February 2015)

Although this interaction seems superficially mundane, it was a surprise for me, a native to Chicago and specifically the South Side, to see these women immediately speaking to each other in such a friendly way. Chicago is known to be one of the most segregated cities in the United States (Moore 2017), and I knew that these two women came from areas that, despite both being working class, rarely interact and have had an open hostility toward one another.

While acknowledging that my own biases were coming into play, I took note that it was a shock to me that these women were able to develop a weak, superficial tie that led to the sharing of information and resources that could help the black woman in her interest to acquire chickens, including pointing her toward other resources. These two women were connecting over the actual practical process of getting chickens: understanding breeds, considering weather, and finding other resources like the feed store for continued troubleshooting. After their interaction, the white woman shared her contact information and offered future help. Although I did not find out if these two women reconnected in the future, I was able to witness the formation of weak ties (Granovetter 1983), which can be tangibly used as a future resource. I call the summation of these ties *practitioner networks* or *communities of practice*, defined as individuals with a variety of weak ties formed specifically to solve problems of subsistence food production *practices*.

I happened upon another interaction as I was waiting to speak to the people at the honey bee table. Staffing the table were two young white people, a man and a woman. After I found out they were from a fashionable North Side neighborhood, I had a sense that they were upper or upper-middle class. They brought some beekeeping materials: a sample hive (without bees in it), honey extraction tools, a beekeeper's helmet and gloves. An older black man walked up to them and started asking trouble-shooting questions. Although I could not decipher the man's class and geography fully, from his questioning it seemed to me that he was urban and working to middle class:

BLACK MAN: So, my bees, they aren't doing well this Winter. They keep dropping off.
WHITE WOMAN: I've been hearing this a lot. Since we had that couple of warm days and now it got so cold recently, this kind of fluctuation is not good for the bees.
M: Okay, okay. So you know what I can do about it?
WW: Ah, so there are a couple of things I've heard of people doing. When it gets warm you can switch them to a Styrofoam hive. Or you can cover the hive in quilting or tar paper on really cold days. But you could have a number of other problems. On a warmer day, check their honey supply, check for a mite infestation, and check to see any mice haven't set up shop inside their hive.

(Field notes 14 February 2015)

The conversation continued with specific ways to identify what might be killing the man's bees this winter, and ways to eliminate what might be the cause and then solutions. This conversation lasted at least 20 minutes, and was interspersed with a lot of jokes and laughter, as well as the exchanging of information for future contact. The interaction between these people is again, unusual for me to see, with my pre-formed biases as a Chicagoan. Certainly, there is not an active hostility to what I assume are liberal-leaning young upper-class urbanites and a lower-class person of color, but it is rare to even imagine the social spaces in which these individuals would have the opportunity to interact. Not only were they increasing social capital (Putnam 2000) through sharing information, but they were also potentially developing weak ties (Granovetter 1983) to be mobilized in the future.

Another similar free event called the Chicago Chicken Coop Tour, hosted by the Chicago Chicken Enthusiasts, had similar results. In this event, chicken owners across the city opened up their private spaces to share information and methods for keeping chickens and other poultry.[3] Again, I noted connections across diverse dimensions of difference on practice-related topics similar to the expo topics, but with a heavier focus on chicken coop construction and weather/predator protection.

An interaction from my field notes that stood out was between an older lower-class black woman from Chicago's south side who was a host on the tour, and a mid-thirties upper-class white woman from a predominantly white neighborhood on Chicago's southwest side. The black woman was showing visitors her coop, which was housed beneath her already existing deck, and was outfitted with a piece of plywood as a door. Her coop design likely cost under $100 for construction. The white woman commented on how she ordered her coop online and it cost her over $1,000:

WHITE WOMAN: Oh my gosh. This set up is incredible! And so simple. And so cost-effective. Why didn't I think of this? Oh, my husband is going to kill me if he finds out that this was an option. You know, a neighbor of mine wants to set up a coop but all I knew was about these expensive coops like the one I bought. I've got to tell her about your set up! So, how did you do the nesting boxes, can I crawl under there and check it out?

BLACK WOMAN: Of course! I don't want to rub it in, but the chickens don't care if it's fancy. They need some warmth, food, water, protection. They got all they need right under here.

(Field notes 19 September 2015)

This interaction is among individuals that differ in age, race, and class, but share gender and urban geography. The geographic similarity has made it so they can problem solve on their shared experiences of limited space for keeping chickens in an urban setting. Certainly, I cannot make claims about the persistence of this social tie. Yet, what the data does suggest is that, for the purposes of this interaction, these women are sharing strategies of practice through this tie that can provide new information on the practical aspects of keeping livestock in a city, and a feeling of pride and efficacy in sharing with the other.

In addition to these moments I witnessed during ethnography at one-time events wherein producers shared resources on practices, interview data suggest that SFPers are also forming lasting communities of practice. Michael and Margot are a middle-class white suburban couple that keep a large vegetable garden, fruit trees, and chickens. During our interview, Michael explained:

The guy who lives behind us, he's from the Caribbean. He gardens and we met him because he walked by and saw we were turning up the dirt. So, he came by and gave us some pointers and said he would bring some chicken poop for us. *Without the garden, we would never have met him.* [Emphasis added.]

Margot explained that they now trade strategies with their neighbor for gardening in an ongoing relationship. Here the results suggest that not only did Michael and Margot form a weak tie with their neighbor whom they may not have otherwise met over the topic of practice, they have mobilized this resource for its utilitarian value and the relationship persists over time.

Lucia described a similar phenomenon with her neighbors. Lucia is Latina, and reported that she did not know how to plant, harvest or even cook certain vegetables that were out of her cultural repertoire until she made a connection with an older black woman from the neighborhood, Dotty:

> Yeah, people come by and ask questions. The older folks in the neighborhood give us tips [on gardening]. I never knew about how to grow and cook collard greens until my neighbor Dotty came by with some seeds and then showed me what to do with them when they came up.

The result is similar across my sample – SFPers find ways to connect through practical problem solving, especially with individuals with whom they may have little else in common. The result, as with the Advocates for Urban Agriculture expo and the chicken coop tour, is both increased civic activity (Putnam 2000) and drawing on weak ties for relevant information and aid (Granovetter 1983). Importantly, connecting over problems of *practice,* rather than attitude or ideology, allows for a much more inclusive coalition and for ties to be formed across greater dimensions of difference. I argue in the next section that these communities of practice are beginning to form the basis for alternative social and economic shadow structures that can be utilized to provide pragmatic solutions to perceived social problems and risk (Polanyi 1944).

Several participants reported that a new or renewed interested in self-producing food has strengthened already strong ties (e.g. with family) because this was a shared practical interest. Margot explained how taking part in SFP has improved her personal relationship with her father, which had been strained and distant due to political differences in recent years:

Margot: My dad's political views are very different from ours. So, [gardening is] the common ground.
Michael: And her father thinks [gardening] is a more true thing in life than those other political things.

Similarly to the majority of the sample, Michael and Margot suggest that it is easier to connect over matters of practice rather than discussions of attitude and identity, especially in the politically polarized climate of modern America.

Even though George, a black man who lives in a rural area, is in his late sixties, he describes reaching out to his older relatives, some of whom still live in the South, "When we first got chickens I was all the time on the phone with my Aunt Shirley...we didn't talk too much before then." Here we see two examples of strengthening what were once weakening ties with family members, which typically represent strong ties. If, as Granovetter suggests, "strong ties have greater motivation to be of assistance and are typically more easily available" (1983, 209), we can see that it is easy for Margot and George to reliably draw on strong ties to gain information, thereby strengthening the relationship further.

Increased community connection has been shown in some studies of alternative food networks among homogenous communities, such as immigrants (Gray 2014; Mares and Pena 2011; Minkoff-Zern 2014), communities of color (Block et al. 2011; Gottlieb and Joshi 2010; Morales 2011) as well as privileged communities (Guthman 2008; Johnston 2008). I find through the act of SFP, people are crossing social boundaries to connect with others on issues of *practice*. The data reveal that despite initially taking part in SFP as privatized solution to a perceived social problem, producers are also finding important social and community connections. As we will see in the next section, the formation of these social ties has sometimes led to the development of shadow structures that could provide templates for resilient, post-capitalist futures.

3. Shadow structure development

There are two important types of shadow structure development that result from the social ties described in the last section: 1) the potential for *political* mobilization of these weak ties, and 2) the opportunity to engage in alternative *economic* exchanges.

First, the existence of these various weak and strong ties within the practitioner networks provides an ever-strengthening network of people ready to mobilize *politically* on issues they care about, specifically on issues related to the ability to take part in SFP. Several participants described cases of utilizing this horizontal practitioner network to support the right to keep chickens. Noah and Joann are a white working-class suburban couple with three small kids who keep ducks, vegetables, and fruit trees, and barter for meat. Noah describes a moment in 2007 when the City of Chicago, which had a long-standing ordinance allowing backyard chickens, began considering banning poultry:

So, in 2007 the alderman for the neighborhood just north of us here, she hates backyard chickens. *Hates* them. And that's okay, some people really think it's not acceptable. For whatever reason. But she hates them so much she wanted to pass a city-wide ban on backyard chickens. And people came out of the woodwork they got organized, that's how the Chicagoland Chicken Coop Tour got started was back in '07. To show that [keeping chickens] wasn't this horrible thing.

Several other urban and suburban SFPers described similar stories of threats to laws protecting the right to keep animals. In each case, SFPers describe mobilizing with their communities of practice to combat the perceived threat.

In this particular case, the mobilization did not take the form of protests, demonstrations or social movement tactics like road blocking (Almeida and Chase-Dunn 2018), but developing a more formal space to share resources, organizing a tour to demonstrate the benefits of keeping chickens, and communicating with neighbors to show support for the right to keep chickens. A group called the Chicago Chicken Enthusiasts formed and began to share resources through such means as a crowd sourced website with resources on feed, supplies, housing, eggs, end of life, as well as recommendations for websites, classes, and stores.

The website also contains a policy advocacy page dedicated to poultry policy in and around Chicago and examples from other cities, handouts on the benefits of keeping poultry and common misconceptions, and how to exercise your individual rights to keep poultry (Chicago Chicken Enthusiasts 2018). In Noah's description, there was a multi-pronged mobilization, with both the formation of this group and political resources, but also a campaign to change public perception by hosting the chicken coop tour. This is a very different kind of political mobilization, one that emphasizes sharing knowledge, empowering individuals, and changing perception through experiential learning about new practices.

In Noah's view, keeping chickens is something that has been done in urban areas, especially among low-income communities, for generations:

It was just normal everyday stuff. 'Cause it was normal everyday stuff going way back, right? And some people just kept doing it and maintained this tradition and the city never changed its ordinance and when this alderman [city council member] tried to change it all these people came out and said 'No, no we have to keep this. This makes our city special, it's unique.' And so in

Chicago there's no ordinance about raising any sort of livestock as long as it's not for meat production. And that means you don't do backyard slaughter, you don't raise animals and churn through them just for the purpose of taking them to a slaughter house. That's prohibited, so meat production's prohibited except for like commercial properties that are zoned for that. But otherwise you can raise chickens, you can raise ducks, you can raise peacocks, you can raise guinea-fowl. There are people with goats in the city of Chicago. They raise them just for milk.

There is an important distinction, then, between the current study and studies of sustainable consumption like Szasz's work on inverted quarantine through green consumption (2007) wherein people lose interest in political action because of a false sense of security. In this case, when there are threats to this alternative mode of production, SFPers will mobilize to protect their rights through their horizontal networks (Evans 2008) of fellow practitioners, many of whom they have been forced to reach out to in order to solve problems of practice.

This is a perfect example of what Wright (2010) describes in his work: a moment wherein interstitial strategies (keeping livestock in urban and suburban areas) are challenged by the larger capitalist structure (city council threatening ordinances) and new interstitial strategies result (resistance to city council, development of more resources, changing local public perception of practice through formation of tours). Just as Wright describes, "there will thus be a kind of cycle of extension of social empowerment and stagnation as successive limits are encountered and eroded" (2010, 235). In the case of both the individualized decision to keep chickens, and the resultant push to protect the ability to do so, we see individuals politically motivated by the desire to protect a *practice*, rather than forming groups based on attitude or ideology.

One of the results of this pushback from the city councilperson was to develop the chicken coop tour, described earlier in the chapter, all over the city wherein individuals with chickens opened their home to visitors for free. The intended result of the chicken coop tour was to demystify chickens so that they were not politically targeted. Yet, there was an unintended consequence that was reported by several of my participants: it helped some people decide to keep their own chickens. Jenny, a white urban middle class woman, actually decided it was possible for her to keep chickens of her own after attending the Chicago Chicken Coop Tour for multiple years:

My son and I have gone on the Windy City Coop Tour for a couple of years and so we bought the house...I was like 'Oh my God, my dreams are coming true!' So I've joked for a while that my life-long dream is to own chickens so we got started on that first.

Along with a minority of my sample, like Madeline, Jenny absolutely sees this as a political action, "You know, it's getting hot, things are going to get, there's going to be a lot of food scarcity soon why don't we learn how to make some of our food?"

Jenny explains that it was the contact she had with the Coop Tour that helped her to see what was possible in an urban setting, "I think what was empowering was...on the Coop Tour I was surprised just to see what people are doing with their spaces." While the choice to start raising chickens does not seem particularly political considering our current conceptions of political action, Jenny's choice is a result of a concerted effort by a hidden community to promote an alternative kind of livelihood, one that is particularly under threat by the neoliberal agenda. That is, as a direct result of a threat to practice (city councilman threatening to change the ordinance) there was a concerted effort made among this community of practice to expose the larger community to the "normalcy" of this set of practices (keeping livestock in the city), which then unintentionally led to more individuals choosing to take part in the practice. This is a kind of virtuous cycle of sharing resources, building horizontal practitioner networks, and building infrastructure to produce food. It represents quite well a shadow structure as it is not in direct conflict with the larger social or economic system, but exists in parallel to it and is building the social, economic and practical infrastructure that can provide for an entirely different kind of system if scaled up and strengthened.

It is noteworthy that this means of political action – sharing resources, educating and raising awareness in the community, and arming practitioners with the knowledge of their rights under obscure city or municipal policy – differs quite substantially from the kind of political action of protest or demonstration that is commonly the focus of social movement scholars, even those within the Polanyian framework (Burawoy 2017; Evans 2000; Evans 2008). This confirms the work of Galt, Gray and Hurley (2014) and Holt-Gimenez (2011) that suggests these communities of practice can potentially be sites for political change around issues of concern for this network, thereby protecting their continued development.

At this point it is important to ask: how do we define a social movement in the modern era? Sociologist Carolyn Lee's edited volume *Democratizing*

Inequalities (2015) puts forth several important studies which find surprising ways in which public participation in recent social movements can have significant negative social outcomes. Within the volume, studies demonstrate that participation in public social movement activity can fiscally benefit corporations (Lee, McNulty, Shaffer 2015; Walker 2015), put more authority in the hands of government and corporations (McQuarrie 2015; Kreiss; Schleifer and Panofsky 2015), and paradoxically can exacerbate inequality while promoting the ideal of democratic participation through technology and social media (Eliasoph 2015; Meyer and Pullum 2015; Polletta 2015). The takeaway message from this set of research is twofold: disadvantaged populations are unable to take part in participatory or public politics, and among the advantaged groups that do take part in politics many are co-opted or controlled for the benefit of those in power (Lee, McQuarrie and Walker 2015).

Here I make the case although these actions and small-scale organizations may not be evidence of a full social movement as generally understood through such actions as protests or demonstrations, this phenomenon is clearly a political shadow structure. This is because it provides an outlet for political action that follows alternative logic to that of mainstream accepted political action. In the case of a threat to their ability to continue to keep animals for food production within the City of Chicago, SFPers mobilized horizontal networks through different modes of political action (education, resource sharing, raising awareness among non-practitioners) that were developed from myriad weak ties (Granovetter 1983).

I argue that the organization around shared behavior is important because it allows for a different kind of community to form. According to Wright, "Forms of social empowerment are likely to be much more durable and to become more deeply institutionalized, and thus harder to reverse, when, in one way or another, they also serve some important interests of dominant groups, solve real problems faced by the system as a whole" (2009, 240). The right to keep chickens, for example, solves the real problem of access to food and therefore creates a durable community that is willing to protect that right.

It has repeatedly been shown by scholars of social movements that participation in social movements is usually centered around shared identity (e.g. environmentalist), and that typically, those that take part in social movements are quite homogenous in demographic characteristics (McPherson, Smith-Lovin and Cook 2001). Yet, in the case of SFPers, data suggest that connections can arise across dimensions of social difference due to a shared interest in practice. This upends the logic of most social movement organizations that often recruit on identity.

The result is the potential for a diverse set of people with group inclusion criteria based on shared behavior rather than class, race, gender or political attitude. I argue that the way these practitioner networks form and organize is so distinct from other modes of political organization and action – because their inclusion criteria are based on practice rather than identity and because they share resources horizontally through direct contact with other practitioners – that they can be considered political shadow structures (Berman 2017) that fit under the umbrella of a Polanyian double movement (1944).

This finding substantiates Evans' claim that modern social movements are likely to be acentered rather than hierarchical (2000; 2008). The data suggest that as commonly accepted modes of political action lose perceived efficacy (Lee 2015; McQuarrie 2015; Kreiss 2015; Schleifer and Panofsky 2015), individuals are turning to actions that are at first a privatized, neoliberal solution (Johnston 2008; Szasz 2007) of self-production. Yet, the development of communities of practice leads to a sort of political shadow structure that does not resemble mainstream modes of "public" action, but allows for swift mobilization for political change on matters relevant to this population.

The second important shadow structure development of this community of practice is alternative *economic* affiliations that do not follow the formal structures of neoliberal capitalism. In one example from my sample, Chad and Brian, a wealthy urban white couple discussed in earlier chapters, describe the way their network is mobilized to share resources:

Interviewer: You give the rabbit compost to community gardens?
Chad: To Maria Elena. Yeah.
Interviewer: And she shares it with different people?
Brian: Yeah, I mean we don't follow the trail of where it goes but we load up – we literally give her a tub or a couple tubs at a time. She has a set of tubs and we have a set of tubs and we rotate it.

Chad and Brian have cultivated many informal working relationships with people they have met through networks of food producers. In this case, they share their excess compost with a woman who is connected to several community gardens, who then distributes this resource that is critical to productive gardening to these many other SFPers.

Although it seems quotidian, this is a physical example of a concept central to modern environmental Marxism, as well as one important characteristic in Weber's description of pre-modern society: closing the metabolic rift (as discussed in detail in Chapter 3) (Foster 1999;

Foster and Holleman 2012). The idea is that central to industrial society is the linear movement of waste from extraction, production and disposal. In other words, the waste does not go back into the site of extraction in order to replenish that ecosystem, instead it ends up in a different space (for waste disposal, like a landfill) where it is concentrated with other waste and therefore cannot add back the nutrients to the Earth.

In pre-modern society, waste cycles were cyclical, often because the scale was smaller. When waste was produced, small-scale producers found some way for this waste to be re-used that would add benefits to an ecosystem (Foster and Holleman 2012, see Table 3.1 in Chapter 3). Here Chad and Brian are doing exactly this through this alternative social arrangement. They have found someone who can take their animal waste and use it to create compost for community gardens, thereby closing the nutrient cycle, rather than putting that waste into a landfill.

I argue that this is evidence of an economic shadow structure (and another kind of Polanyian double movement) because it is relying on a connection that is entirely parallel to the formal economy and shares critical resources through horizontal networks. In this case, it is beneficial for Chad and Brian to have a place to dispose of their excess animal waste. For the gardeners receiving this waste, it is incredibly valuable in improving their own food production, without added cost. It is an economic shadow structure that involves no payment and no formal records, that is mutually beneficial and mirrors pre-modern social organization (Foster 1999; Foster and Holleman 2012; Marx 1990; Weber 1930).

In another such relationship, Chad and Brian describe how a social tie evolved with a local farmer:

Brian: We also are part of a CSA [community supported agriculture] membership and at this point it's mainly because we we've been with them for so long before we ever had a farm we were with this farmer and he did such a great job and we love supporting him.

Chad: Now he's a friend of ours. He came to our anniversary party. In fact [his wife] is the one that has like told me that if both of those clutches of eggs wind up hatching [and we have too many ducks as a result] that's where they'd go.

Brian: Yeah. We've shared turkeys with them. We've shared ducks with them. Basically if we get more than we can handle it goes –

Chad: They're our overflow.

Brian: And then we split like the bounty so like they'll raise it and they'll give us half back. It's a sort of process.

Here Chad and Brian have developed a seemingly simple, mutually beneficial social tie (Granovetter 1983) to deal with their problems of space constraints raising livestock in the city. When they have too many eggs hatched, they have a place to bring their animals so that they can continue to live. In return for raising it, the farmer gets half of the meat and Chad and Brian get the other half.

The significance of this kind of economic relationship should not be understated. Their exchange of goods and services is not being recorded by any economic measures, as it is completely lacking any monetary exchange. The arrangement is mutually beneficial for both the farmer and Chad and Brian, and results in a strong personal connection, as evidenced by the farmer's presence at their anniversary party, thereby benefitting the growth of civil society (Granovetter 1983; Putnam 2000). In both of these examples, Chad and Brian have formed Gemeinschaft-like social ties because of shared interest in the practice of SFP that help them to solve some of the issues that arise from raising animals for food, mirroring pre-modern social structures (Foster and Holleman 2012). These economic shadow structures are an example of an active cultivation of alternatives to mainstream, in-dustrial methods of production and consumption. These particular types of environmental, social, and economic shadow structures have potentially transformative implications.

Drawing from Table 3.1 in Chapter 3 on "Weber's characteristics of modern and pre-modern eras," we can see that these two relationships described by Chad and Brian have several important similarities to pre-modern societies: organic limitations to production, common (or shared) land use, business (or production) in home, closed nutrient cycle, and enchantment (drawn from Foster and Holleman 2012). I argue, similarly to Berman (1981; 2017), that because pre-modern society was once structured in this way, it is within the realm of pos-sibility to develop shadow structures that draw from these pre-modern practices that would be successful in providing viable alternative ways of living to the hegemonic system of late capitalism (Burawoy 2017; Evans 2000; Harvey 2017; Polanyi 1944; Wright 2010). But I also acknowledge that the development of these alternatives is rife with struggle and contradiction (Galt et al. 2014; McClintock 2014), such as the promotion of individualized solutions (Szasz 2007), which can represent a neoliberal logic, explored in the first section of results.

Nearly every participant described the development of similar kinds of relationships of bartering, sharing, or exchange (of either goods or information), with differing levels of commitment. Emma is a working-class white suburban woman who keeps chickens, hunts, fishes, grows vegetables, forages, gleans, has an orchard, and barters. She reported that she produces over 90 percent of her own food year-round, especially through the use of food preservation. As a part of her food growing, there is often a surplus that she does not have the ability to preserve. When she has this surplus, she finds ways to re-distribute it so that it does not go to waste.

Emma explained that she is constantly sharing surplus to food banks and neighbors, and that it results in more social connection, "Nowadays we often don't see our neighbors. So, going by in the afternoon and dropping off a half dozen eggs…it really promotes a sense of community and neighborliness." Here, again, is evidence of the development of this Gemeinschaft community connection, thereby strengthening civil society (Putnam 2000) through the development of weak ties (Granovetter 1983), as well as contribution to the gift or sharing economy, which is an example of an economic shadow structure (Berman 2017).

Madeline, the lower-class urban black woman mentioned earlier, sees subsistence food production as a way to combat the economic warfare that has been waged on her South Side Chicago community:

> One year we tried a CSA [community supported agriculture] that delivered to families in Lincoln Park [a wealthy Chicago neighborhood]. Where's the benevolence in that? These people can afford Whole Foods. Staying small and not selling it is political to me. Who am I doing this for? *We don't want the money, we want the garlic.* We want to give it away to the community.

This economy of growing and sharing upends the traditional logic of financial accumulation and growth. Following the logic of neoliberal capitalism, the main goal is to make as much money as possible. However, in this case, growing healthy food in a poor South Side community to sell in a wealthy North Side community to make a small amount of money is less valuable, to Madeline, than the prospect of keeping the healthy food within the South Side, where they need it most, and giving it away for free. What Madeline describes here, I argue, is an economic shadow structure in which members of her community work together to obtain this essential resource outside of the rationality of formal neoliberal capitalism.

Subsistence food producers are making connections in uncommon ways as they are disrupting accepted ways of social, economic or political activity. The result is significantly different modes of social organization, or shadow structures, that can challenge the logic of late capitalism. These economic shadow structures also provide a form of social organization to turn to in case of more catastrophic social collapse (Harvey 2017) by providing for needs individually or on a small-scale in Polanyian double movement activity (1944). The initial decision to take part in SFP for my sample is motivated by the neoliberal logic of enacting private solutions to perceived public problems. Yet, as SFPers get more involved in the practice of SFP, they reach out to others, making surprising connections. The data suggest that these networks of practice have then been mobilized to protect the right to partake in SFP and also help to develop alternative environmental, social, economic or political interactions or patterns of behavior that exist parallel to mainstream social modes of organization.

Seeds of shadow structures

The results of this chapter reveal a similar story to the outcomes of the previous chapters: the processes involved in the development of sub-sistence food production are paradoxical and part of an emerging dialectic, which I argue is a response to some of the failures of late capitalism (Burawoy 2017; Harvey 2017; Polanyi 1944) and resultant looming crises (Beling et al. 2017; Fischer-Kowalski and Haberl 2007; Roberts 2009; Leonard 2011). At once, SFPers are validating neo-liberal logic through enacting individualized solutions while also producing radical alternative shadow structures (Berman 2017; McClintock 2014) both through their own relation to the means of production as well as newly developing communities of practice.

Although the meanings around the initial motivation to self-produce food are often representative of the neoliberal logic of privatized solu-tions (Johnston 2008; Szasz 2007), the data suggest that the behavior of self-producing food often leads to more significant social or political outcomes than simple shifts in consumption patterns. Producers report the need to reach out to human resources to problem solve the myriad difficulties that arise with home food production. This then leads to the formation of communities of practice, that have been studied in some detail among immigrant communities (Gray et al. 2014; Mares and Pena 2011; Minkoff-Zern 2014), communities of color (Block et al. 2011; Gottlieb and Joshi 2010; Morales 2011) as well as privileged commu-nities (Guthman 2008; Johnston 2008). I add to this literature by de-monstrating that individuals are moving across diverse social boundaries to connect with others on topics related to SFP practices.

Two outcomes result from this network formation. First is the development of a diverse community of practice, which is horizontal in structure (Evans 2008). The myriad relationships that make up this community have the manifest function of sharing resources in order to successfully produce food, an important outcome for these individuals who have identified subsistence food production as a meaningful and pragmatic solution to their perceived problems (Kennedy et al. 2017; McAdam et al. 2001). The latent function of these communities is the development of strong and weak social ties (Granovetter 1983) that support civil society (Putnam 2000) and can be mobilized for social change around issues of importance to the community, as Galt (2014) and Holt-Gimenez (2011) have also suggested. I argue that this is a form of political shadow structure, as it represents political organization that is significantly divergent from formal accepted modes of public participation such as protests (McAdam 2000; McCarthy 2013; McCarthy and Zald 1977), but nonetheless achieve political aims through alternate means (Berman 2017; Polanyi 1944).

Second is the development of a set of alternative economic relationships that do not explicitly interact with mainstream capitalist modes of organization. This is an example of what Gibson-Graham calls sites of "new economic becomings" (2006, 77) and what Wright calls "interstitial spaces" (2010). Many of these shadow structures closely mirror pre-modern societies (see Table 3.1 in Chapter 3), which suggests that these types of social organization have been successful in the past, and could be easily re-instated if the contradictions of capitalism come to pass in a more comprehensive way (Beck 1992; Harvey 2017; Polanyi 1944). Networks of subsistence food producers are part of a horizontal network that is not intentionally forming for any explicit social or political purposes according to participants, but has been shown to be able to be utilized as a pragmatic solution to social problems arising from the problems of late capitalism (Beling et al. 2017; Kennedy et al. 2017).

The overall social or political impact of these shadow structure formations is unclear and beyond the scope of this study. Certainly, the potential for social change as these shadow structures develop is staggering, at least according to Berman and others who argue the collapse of capitalism and emergent alternatives of shadow structures is likely to be "the central story of the rest of the 21st Century" (2017, 1). Future research should attempt to quantify the economic, environmental, health, social or political outcomes of taking part in SFP, or other shadow structures. This exploratory research instead is attempting to uncover the processes and meanings involved in the work of shadow structure development by subsistence food producers, a highly understudied group.

What *is* clear from the current study, however, is that the growth of shadow structures (Berman 2017) as the crises of capitalism unfold (Harvey 2017) is paradoxical. Results suggest that certain strains of neoliberal logic and social stratification remain potent forces.

Simultaneously, SFPers are pursuing potentially radical social, economic and political restructuring that could provide alternative modes of social organization as conventional structures fail to provide for needs (Berman 2017; McClintock 2014; Harvey 2017).

Overall, the data suggest that these communities and practices are doing the work of "making people's lives better within the locality, increasing their social ties to each other, and empowering them as they build self-reliant communities and new spaces for social change" (Galt et al. 2014, 137). However, this is just a first step in a "both/and strategy":

> one that recognizes the importance and contributions of…the community economies that they are building, and one that works on the larger political project of recreating a political discourse and political reality in which collective action and state provisioning of basic human needs become revalorized.
>
> (Galt et al. 2014, 141)

This study suggests that those who are interested in responses to documented social problems may want to inquire into these communities of practice that are developing shadow structures that may well ease the transition into a new social era.

Notes

1 I noted between 90 and 100 participants and attempted to speak to as many participants as possible, reaching close to half. In my ethnographic observations I encountered vast diversity in geography (among those I spoke with directly), as well as age, race and gender.
2 In my ethnographic observations, I used place of residence (when I asked participants this question) as a proxy indicator for class. In Chicago, neighborhoods and suburbs are clearly defined (segregated) and are often inhabited by homogenous populations both in terms of race and class. If I found out place of residence, I could make an educated guess about proximate socioeconomic status.
3 I attended this tour in 2015 and 2016 and visited nearly all sites over two days (20–25 sites) each year. The host sites tended to be white and more affluent, yet there were several sites of low income and/or non-white households.

Works cited

Almeida, Paul and Chris Chase-Dunn. 2018. "Globalization and Social Movements." *Annual Review of Sociology*, 44: 1.1–1.23.
Beck, Ulrich. 2016. *The Metamorphosis of the World: How Climate Change is Transforming our Concept of the World*. New York: Polity Press.

Beling, Adrian, Julien Vanhulst, Federico Dermaria and Jerome Pelenc. 2017. "Discursive Synergies for a 'Great Transformation' towards Sustainability: Pragmatic Contributions to a Necessary Dialogue between Human Development, Degrowth and Buen Vivir." *Ecological Economics.* doi:10. 1016/j.ecolecon.2017.08.025.

Berman, Morris. 2017. "Dual Process: The Only Game in Town." In *Are We There Yet?* Brattleboro, VT: Echo Point Books. Essay #27.

Berman, Morris. 1981. *The Reenchantment of the World.* Ithaca, NY: Cornell University Press.

Block, Daniel R., Noel Chavez and Erika Allen. 2011. "Food Sovereignty, Urban Food Access, and Food Activism: Contemplating the Connections through Examples from Chicago." *Agriculture and Human Values*, 29(2): 203–215.

Brunsma, David, Divid Diverfelt and Steven Picou. 2007. *The Sociology of Katrina: Perspectives on a Modern Catastrophe.* New York: Rowman and Littefield.

Burawoy Michael. 2017. Social Movements in the Neoliberal Age. pp. 21–35. In *Southern Resistance in Critical Perspective*, eds. M. Paret, C. Runciman, L. Sinwell. New York: Routledge.

Chicago Chicken Enthusiasts. 2018. "Welcome!" Retrieved June 1, 2018 from https://sites.google.com/site/chicagochickenenthusi/home.

Eliasoph, Nina. 2015. "Spiral of perpetual potential: How empowerment projects' noble missions tangle in everyday interaction." In *Democratizing Inequalities: Dilemmas of the New Public Participation.* Ed. Caroline Lee New York: NYU Press.

Evans, Peter. 2000. "Fighting Marginalization with Transnational Networks: Counter-hegemonic Globalization." *Contemporary Sociology*, 29(1): 230–241.

Evans, Peter. 2008. "Is an Alternative Globalization Possible?" *Politics and Society*, 36(2): 271–305.

Fischer-Kowalski, M. and H. Haberl. 2007. *Socioecological Transitions and Social Change: Trajectories of Social Metabolism and Land Use.* In "Advances in Ecological Economics," series editor: Jeroen van den Bergh. Cheltenham, UK and Northampton, USA: Edward Elgar.

Foster, John Bellamy. 1999. "Marx's Theory of Metabolic Rift: Classical Foundations for Environmental Sociology." *The American Journal of Sociology*, 105(2): 366–405.

Foster, John Bellamy and Hannah Holleman. 2012. "Weber and the Environment: Classical Foundations for a Post-exemptionalist Sociology." *American Journal of Sociology*, 117(6): 1625–1673.

Fussell, Elizabeth. 2008. "Leaving New Orleans, Again." *Traumatology*, 14(4): 63–66.

Galt, Ryan E., Leslie C. Gray, and Patrick Hurley. 2014. "Subversive and Interstitial Food Spaces: Transforming Selves, Societies, and Society–Environment Relations through Urban Agriculture and Foraging." *Local Environment*, 19(2): 133–146.

Gibson-Graham, J.K. 2006. *A Postcapitalist Politics.* Minneapolis: University of Minnesota Press.

Gottlieb, R. and A. Joshi. 2010. *Food Justice.* Cambridge, MA: MIT Press.

Granovetter, Mark. 1983. "The Strength of Weak Ties." *Sociological Theory.* Volume 1. pp. 201–233.

Gray, L. et al. 2014. "Can Home Gardens Scale Up into Movements for Social Change? The Role of Home Gardens in Providing Food Security and Community Change in San Jose, California." *Local Environment*, 19(2): 187–203.

Guthman, Julie. 2008. "Neoliberalism and the Making of Food Politics in California." *Geoforum*, 39(3): 1171–1183.

Harvey, David. 2017. *Marx, Capital, and the Madness of Economic Reason.* New York: Oxford University Press.

Holt-Gimenez, E., ed., 2011. *Food Movements Unite! Strategies to Transform our Food Systems.* Oakland: Food First Books.

Johnston, Josee. 2008. "The Citizen-Consumer Hybrid: Ideological Tensions and the Case of Whole Foods Market." *Theory and Society*, 37(3): 229–270.

Kennedy, E.H., J. Johnston and J. Parkins. 2017. "Small-p politics: How Pleasurable, Convivial, and Pragmatic Political Ideals Influence Engagement in Eat-Local Initiatives." *British Journal of Sociology.* doi:10.1111/1468-4446.12298.

Kennedy, Emily H. and Josee Johnston. 2014. "Social Movements and the Citizen-Consumer: Evidence from the Canadian Sustainable Food Movement." *International Sociological Association World Congress of Sociology*, Yokohama, Japan.

King, Christine A. 2008. "Community Resilience and Contemporary Agri-ecological Systems: Reconnecting People and Food, and People with People." *Systems Research and Behavioral Science*, 25: 111–124.

Kreiss, Daniel. 2015. "Structuring Electoral Participation: The Formalization of Democratic New Media Campaigning, 2000–2008." In *Democratizing Inequalities: Dilemmas of the New Public Participation.* Ed. Caroline Lee. New York: NYU Press.

Lee, Caroline (ed.). 2015. *Democratizing Inequalities: Dilemmas of the New Public Participation.* New York: NYU Press.

Lee, Caroline, Kelly McNulty and Sarah Shaffer. 2015. "Civic-izing Markets: Selling Social Profits in Public Deliberation." In *Democratizing Inequalities: Dilemmas of the New Public Participation.* Ed. Caroline Lee. New York: NYU Press.

Lee, Caroline, Michael McQuarrie and Edward Walker. 2015. "Realizing the Promise of Public Participation in the Age of Inequality." In *Democratizing Inequalities: Dilemmas of the New Public Participation.* Ed. Caroline Lee New York: NYU Press.

Leonard, Annie. 2011. "Global Change: By Disaster or by Design?" Presentation at Tulane University, New Orleans, LA. October 3, 2011.

MacKendrick, Norah and Lindsay M. Stevens. 2016. "'Taking back a little bit of control': Managing the Contaminated Body Through Consumption." *Sociological Forum*, 31(2): 310–329.

Mares, Teresa M. and Devon G. Pena. 2011. "Environmental and Food Justice: Toward Local, Slow and Deep Food Systems." pp. 197–219. In *Cultivating Food Justice: Race, Class and Sustainability*. Eds. Alison Hope Alkon and Julian Agyeman. Cambridge, MA: MIT Press.

Marx, Karl. 1990 [1867]. *Capital: Volume I*. Translated by Ben Fowkes. London: Penguin Books.

McAdam, Doug. 2000. "Culture and Social Movements." In *Culture and Politics*. Eds. L. Crothers and C. Lockhart. New York: Palgrave Macmillan.

McAdam, Doug. Sidney Tarrow and Charles Tilly. 2001. *Dynamics of Contention*. In "Cambridge Studies of Contentious Politics". London: Cambridge University Press.

McCarthy, John D. 2013. "Social Movement Sector." In *The Wiley-Blackwell Encyclopedia of Social and Political Movements*. https://doi.org/10.1002/9780470674871.wbespm196.

McCarthy, John D. and Mayer Zald. 1977. "Resource Mobilization and Social Movements: A Partial Theory." *American Journal of Sociology*, 82(6): 1212–1241.

McClintock, Nathan. 2014. "Radical, Reformist, and Garden-Variety Neoliberal: Coming to Terms with Urban Agriculture's Contradictions." *Local Environment*, 19: 147–171.

McPherson, Miller. Lynn Smith-Lovin, and James M. Cook 2001. "Birds of a Feather: Homophily in Social Networks." *Annual Review of Sociology*, 27: 415–444.

McQuarrie, Michael. 2015. "No Contest: Participatory Technologies and the Transformation of Urban Authority." In *Democratizing Inequalities: Dilemmas of the New Public Participation*. Ed. Caroline Lee New York: NYU Press.

Meyer, David S. and Amanda Pullum. 2015. "The Social Movement Society, the Tea Party, and the Democratic Deficit." In *Democratizing Inequalities: Dilemmas of the New Public Participation*. Ed. Caroline Lee. New York: NYU Press.

Minkoff-Zern, L-A. 2014. "Hunger Amidst Plenty: Farmworker Food Insecurity and Coping Strategies in California." *Local Environment*, 19(2): 204–219.

Moore, Natalie. 2017. *The South Side: A Portrait of American Segregation*. Chicago: Picador.

Morales, Alfonso. 2011. "Growing Food and Justice: Dismantling Racism through Sustainable Food Systems." pp. 149–176. In *Cultivating Food Justice: Race, Class and Sustainability*. Eds. Alison Hope Alkon and Julian Agyeman. Cambridge, MA: the MIT Press.

Polanyi, Karl. 1944. *The Great Transformation: The Political and Economic Origins of Our Time*. Boston, MA: Beacon Press.

Putnam, Robert. 2000. *Bowling Alone: The Collapse and Revival of American Community*. New York: Simon and Schuster.

Roberts, Michael R. 2009. "Control Rights and Capital Structure: An Empirical Investigation." *The Journal of Finance*. https://doi.org/10.1111/j.1540-6261.2009.01476.x.

Schleifer, David and Aaron Panofsky. 2015. "Patient, Parent, Advocate, Investor: Entrepreneurial Health Activism from Research to Reimbursement." In *Democratizing Inequalities: Dilemmas of the New Public Participation*. Ed. Caroline Lee. New York: NYU Press.

Szasz, Andrew. 2007. *Shopping our way to safety: How we changed from protecting the environment to protecting ourselves*. Minneapolis: University of Minnesota Press.

Walker, Edward T. 2015. "Legitimating the Corporation the Public Participation." In *Democratizing Inequalities: Dilemmas of the New Public Participation*. Ed. Caroline Lee New York: NYU Press.

Weber, Max. 1930. *The Protestant Ethic and the Spirit of Capitalism*. Translated by Talcott Parsons. New York: Scribner.

Wright, Erik Olin. 2010. "Interstitial Transformations." Chapter 10 in *Envisioning Real Utopias*. London: Verso.

7 Conclusion: "We've got to find a solution"

This is the age of machinery,
A mechanical nightmare,
The wonderful world of technology,
Napalm hydrogen bombs biological warfare,
This is the twentieth century,
But too much aggravation
It's the age of insanity,
What has become of the green pleasant fields of Jerusalem?
Girl we've got to get out of here
We've got to find a solution
I'm a twentieth century man but I don't want to die here.
                         The Kinks, *20th Century Man*

Understanding major historical shifts and their accompanying social problems is central to the discipline of sociology. Marx lamented the rise of capitalism, and the resultant exploitation of the working class (Marx and Engels 1967 [1848]). Weber discussed the development of industrial civilization and suggested this led to a deep disenchantment from material reality (1930; Berman 1981). Polanyi suggested the commodification of nature and human labor would eventually lead to a double movement resistance and the development of a new world system (1944). The work of these theorists of industrial society has been re-interpreted in recent years to reveal a profound critique of human–environment relations in the modern era (Beck 2016; Foster 1999; Foster and Holleman 2012; Urry 2010).

   This book has sought to understand not only social problems, but emergent solutions. This is not a story (solely) about risk and precarity (Beck and Ritter 1992; Burawoy 2017; Polanyi 1944), but about resilience (Hopkins 2014; King 2008). The findings are paradoxical (Galt Gray, and Hurley 2014) and in a dialectical movement (McClintock 2014) from reinforcing capitalist logic to radical countermovement. People who grow their own food, or subsistence food producers (SFPers), are developing new, inclusive and diverse cultures around food production that help to counter the alienation they feel. They do this through the act of physical work (Chapter 4). Because of their proximity to food production, or ecological embeddedness (Whiteman and Cooper 2000), diverse SFPers are also reporting a re-enchantment with nature (Berman 1981; Chapter 5), leading to positive environmental practices. This leads to a new conception of environmentalism, one that is inclusive of those practicing pro-environmental *behaviors*, not just those with environmentalist *identities*.

Though the decision to take part in SFP can be tied up with neoliberal meanings (Johnston 2008; Szasz 2007), the resultant horizontal practitioner networks of producers leads to a) mobilization for political change, and b) the development of alternative economic relationships, which I call shadow structures (Chapter 6). I will review the findings of this book and the implications both for future study as well as activism and social change.

Overview of results

Chapter 4: Who are subsistence food producers in Chicago? Meanings across class of alienation and viscerality

In this first substantive chapter I attempt to understand the makeup of those taking part in subsistence food production in Chicago. When entering my research, I did not know exactly how meanings around SFP would coalesce. Which groups would have shared meanings? Would habitus (Bourdieu 1984) center around shared geography, gender, race, age or class?

In this chapter I describe the surprising and paradoxical (Galt et al. 2014; McClintock 2014) finding that certain meanings around both the act of subsistence food production and the product of it are shared across dimensions of difference including class, race, geography, gender and age. These meanings are two-sided. On the one hand is a sense of alienation from the quotidian aspects of modern life in the era of late capitalism. I use classical sociological theories (Marx 1990; Weber 1930) and their modern reinterpretations (Berman 1981; Foster 1999; Foster and Holleman 2012) to explain how alienation and disenchantment are central to modern industrial life.

On the other hand, I recurrently find in my sample a desire to counter this disenchantment (Berman 1981) through the act of SFP which is perceived as able to bring one closer to work, nature, and community. To explain this, I use Hayes-Conroy's (2011) theoretical argument that in the viscerality, or physicality, of food lies its ability and power to cross social boundaries and unite people. I build on this initial finding in the following two chapters, demonstrating the important ways in which SFP brings people closer to nature and connects them to community, with potentially important social implications.

Chapter 5: Marginalized environmentalism and ecological embeddedness

Drawing on the theoretical framework of ecological embeddedness (Mincyte and Dobernig 2016; Whiteman and Cooper 2000), I find that even though my sample represents a diverse set of people practicing subsistence food production, every participant is engaging in some pro-environmental behavior, such as: composting, cyclical waste cycling, water catchment/recycling, limited chemical inputs/organic pest and disease management, soil remediation, and others. I argue that it is because participants are ecologically embedded (Whiteman and Cooper 2000), or physically close to the material conditions on which they rely to produce food, that they make more conservation-minded decisions in their food growing practices. I find that this is true across the spectrum of SFPers in my sample, whether or not they identify as environmentalists.

The implication of this for future research and activism is twofold. First, environmental researchers and activists must stop looking solely at those holding environmentalist *identities* or *attitudes* as being as the forefront of environmental solutions. Environmentalists have been shown to be some of the worst greenhouse gas emitters in the world (Boucher 2017), as a function of their socioeconomic status. Further, the solutions that have arisen out of the myopic attention placed on self-described environmentalists have advocated for ecological modernization, or a set of solutions that prioritizes slight changes in consumer behavior or the hope that efficiencies in technology will lead to societal transformation, all of which have fallen short in meeting any expectations for the kind of change we need to combat the scale of ecological problems we are facing (Akenji 2014; Geels et al. 2015; Isenhour, Martiskainen and Middlemiss 2019; Lorek and Fuchs 2013; Middlemiss 2018; ORourke and Lollo 2015).

Instead, we must look to those enacting pro-environmental *behaviors.* In other words, it is sometimes those who are at the far margins of the environmental movement (e.g. white rural populations, or urban populations of color), who may be at the forefront of innovating pro-environmental solutions. We must, as scholar activists, have the humility to look to those on the margins for strategies for many of the looming environmental crises we face as a civilization.

Second, researchers must start paying attention to certain behaviors that are often ignored as being a pro-environmental solution. It is a dead-end to focus on fostering pro-environmental *attitudes* as there is a significant gap between how people identify and how they behave (Blake 1999; Hoggett 2013; Kollmuss and Agyeman 2002; Whitmarsh, Seyfang and O'Neill 2011). Instead, we must look not only at obvious positive behaviors like recycling, but strategies at the margins such as dumpster diving, freecycling, and sharing economies such as tool or seed libraries. I would suggest pushing this even further in suggesting that many unnamed but universal subsistence strategies of the poor around the world are at the cutting edge of environmental shadow structures such as: bartering, hunting and fishing, transport sharing, reusing or passing goods, alternative currencies, and more. I will discuss these alternative economic shadow structures in some more detail in the next section.

Chapter 6: Horizontal practitioner networks and post-capitalist shadow structures

As I have demonstrated thus far, the development of this specific alternative to capitalist modes of being, subsistence food production, is full of contradictions and paradoxes. For example, in Chapter 6, I demonstrate the paradoxical way in which the decision to take part in SFP starts out as a neoliberal, individualized solution to perceived social problems (Johnston 2008; Szasz 2007) and eventually becomes a way to strengthen social ties (Granovetter 1983), mobilize for political change, and develop alternative economic arrangements which I call shadow structures (Berman 2017), a type of Polanyian double movement (1944).

The idea of what constitutes a social movement is constantly changing and adapting to the current political and social climate. Scholars have demonstrated that what we imagine as an ideal social movement (like those public protests of the civil rights struggle (McAdam 2000; McCarthy 2013; McCarthy and Zald 1977)) is now no longer possible both because of a heightened police state, the co-optation of public demonstrations, and the lack of political efficacy of such activity (Lee 2015). People are then turning to private solutions for public issues such as the "inverted quarantine" like buying bottled water to address a perceived problem with public water quality (Szasz 2007). I find that the decision to take part in SFP begins as this kind of individualized, neoliberal solution to the perceived problems of alienation, food quality and access.

Paradoxically, because SFP is not simply a shift in consumer action but a production behavior which requires information and problem solving, SFPers report reaching out to others for help. I demonstrate the ways in which this adds to both strong (family, close friends) and weak (acquaintance, e.g., neighbor) ties (Granovetter 1983) to form a Gemeinschaft-like network through which to share resources. I add to the literature by showing how this network has been employed toward the development of both political and economic shadow structures. I argue these shadow structures are a kind of Polanyian double movement meant to diminish risk created by the commodification and extreme fluctuation in value of land, labor and money (1944), while simultaneously providing an alternative set of social arrangements meant to help ease the transition as larger social structures of capitalism fail (Berman 2017).

One example of the formation of a political shadow structure is illustrated in the chapter. Food producers, specifically those interested in keeping livestock within the city limits of Chicago, mobilize in response to a city councilman's threat to change the ordinance for livestock. The community rallies to develop resources for individuals that keep chickens, but develops a public awareness campaign in the form of a free tour of chicken coops. I argue this can be understood as a political shadow structure (Berman 2017), because it is a kind of political organization that exists parallel to mainstream modes of social action like protest or taking part in large-scale social movement organizations. I also argue that these arrangements constitute a shadow structure because they follow a logic different from these other systems that rely on often hierarchical modes of political action centered around shared attitudes (and result in highly homogenous groups (McPherson, Smith-Lovin and Cook 2001)), whereas these are horizontal networks (Evans 2008) centered around a certain behavioral practice leading to a quite diverse population.

Second, several SFPers have developed what I argue are economic shadow structures (Berman 2017), because the way in which they are organizing with others meets presented needs and exists parallel to (not in direct conflict with) the hegemonic economic system. Some of the shadow structures reported by SFPers include alternative economic arrangements that are completely outside of formal modes of exchange, using networks to develop cyclical (rather than linear) waste cycles, or re-distribution of surplus resources to in-need communities.

The implications of these findings are not to be understated but require further study. Although taking part in SFP starts out as a privatized, neoliberal solution, it develops into a radical and transformative shadow structure that could provide an alternative to hegemonic, exploitative social organization of late capitalism. Prominent sociologist John Urry suggests, "the twentieth century has left a bleak legacy for the new century, with a very limited range of possible future scenarios" (2010, 191). Yet, despite this bleak outlook, the creativity of those at the margins endures, and the results of this study suggest it is worth further exploration. The current research attempts to understand just one small community of subsistence food producers who are endeavoring to explore these new beginnings.

Future research and overall contribution

As I have discussed in Chapter 2, the strengths of qualitative research lie in uncovering meanings and processes, not addressing large scale social change. If, as other scholars have suggested, we are on the precipice of another "great transformation" (Polanyi 1944), then we need research to more fully understand the outcomes of these emergent shadow structures. It is important, I argue, to pay attention to marginalized communities and behaviors and to understand these phenomena as *connected* to a larger development of alternatives to late capitalist livelihoods.

We must have interdisciplinary scholarship that dissects the effects of this network of shadow structure development in terms of human health, environmental health (including such indicators as soil health, ecosystem services, or impacts on species decline) as well as attempts to understand the alternative social structures and networks (e.g., horizontal versus vertical). We must also explore the resultant economic activities and emergent political strategies. Equipped with this information, activists can help to both foster already burgeoning shadow structure activities as well as build a coalition of shadow structure practitioners to enact po-litical or economic change. The outlook for the potentially overlapping social, economic, environmental and political crises that loom on the horizon is grim. Yet, human life and struggle through resistance and creativity continues.

The findings of this book offer several important contributions to the literature of sustainable consumption. First, scholars interested in solutions to unfolding environmental crises must look beyond those who hold pro-environmental attitudes and identities, but instead consider those who are enacting pro-environmental behaviors. Beginning to pay attention to these marginalized groups has the po-tential to lead to greater inclusivity and diversity and provides a point of hope for the environmental movement in the resilience being enacted if we can shed our biases about who gets invited into the conversation.

Second, the understanding of alienation and feelings of disenchantment in the modern era that was so eloquently explained by the founders of sociology (Marx 1990; Weber 1930) is still ubiquitous in this era, and is driving unexpectedly diverse individuals to try myriad ways to reconnect to physical work, nature and one another. Finally, the particularities of the resilience and creativity being enacted include social, economic and political shadow structures that upend the logic of capitalism, including mainstream political action, deleterious environmental actions, and alienated social ties. These new parallel forms of social organization mirror many characteristics of pre-modern societies (Table 3.1), and may very well be sowing seeds and forming modes of organization that will inform the post-industrial era.

On the eve of WWII and all its resultant destruction, poet W.H. Auden saw hope in radical social connection:

> Defenseless under the night
> Our world in stupor lies;
> Yet, dotted everywhere,
> Ironic points of light
> Flash out wherever the Just
> Exchange their messages:
> May I, composed like them
> Of Eros and of dust,
> Beleaguered by the same
> Negation and despair,
> Show an affirming flame.
> > 1 September 1939

The results of this book suggest that it is in those "points of light" that we must cast our gaze to and embed our hope in for the coming era.

Works cited

Akenji, L. 2014. "Consumer Scapegoatism and Limits to Green Consumerism." *Journal of Cleaner Production*, 63: 13–23.

Beck, U. and M. Ritter. 1992. *Risk Society: Towards a New Modernity*. London: Sage Publications.

Beck, Ulrich. 2016. *The Metamorphosis of the World: How Climate Change is Transforming our Concept of the World*. New York: Polity Press.

Berman, Morris. 1981. *The Reenchantment of the World*. Ithaca NY: Cornell University Press.

Berman, Morris. 2017. "Dual Process: The Only Game in Town." In *Are We There Yet?* Brattleboro, VT. Echo Point Books. Essay #27.

Blake, J. 1999. "Overcoming the Value–Action Gap in Environmental Policy: Tensions between National Policy and Local Experience." *Local Environment*, 4(3): 257–278.

Boucher, J.L. 2017. "Culture, Carbon and Climate Change: A Class Analysis of Climate Change Belief, Lifestyle Lock-in, and Personal Carbon Footprint." *Socijalna Ekologija*, 25(1): 53–80.

Bourdieu, Pierre. 1984. *Distinction: A Social Critique of The Judgement of Taste*. Cambridge, MA: Harvard University Press.

Burawoy, Michael. 2017. "Social Movements in the Neoliberal Age." pp. 21–35. In *Southern Resistance in Critical Perspective*, eds. M. Paret, C. Runciman and L. Sinwell. New York: Routledge.

Evans, Peter. 2008. "Is an Alternative Globalization Possible?" *Politics and Society*, 36(2): 271–305.

Foster, John Bellamy. 1999. "Marx's Theory of Metabolic Rift: Classical Foundations for Environmental Sociology." *The American Journal of Sociology*, 105(2): 366–405.

Foster, John Bellamy and Hannah Holleman. 2012. "Weber and the Environment: Classical Foundations for a Post-exemptionalist Sociology." *American Journal of Sociology*, 117(6): 1625–1673.

Galt, Ryan E., Leslie C. Gray and Patrick Hurley. 2014. "Subversive and Interstitial Food Spaces: Transforming Selves, Societies, and Society–environment Relations through Urban Agriculture and Foraging." *Local Environment*, 19(2): 133–146.

Geels, F.W., A. McMeekin, J. Mylan, and D. Southerton. 2015. "A Critical Appraisal of Sustainable Consumption and Production Research: The Reformist, Revolutionary and Reconfiguration Positions." *Global Environmental Change*, 34: 1–12.

Granovetter, Mark. 1983. "The Strength of Weak Ties." *Sociological Theory*. Volume 1. pp. 201–233.

Hayes-Conroy, Jessica, 2011. "School Gardens and 'Actually Existing' Neoliberalism." *Humboldt Journal of Social Relations*, 33(1/2): 64–96.

Hoggett, P. 2013. "Climate Change in a Perverse Culture." pp. 56–71. In S. Weintrobe (ed.), *Engaging with Climate Change: Psychoanalytic and Interdisciplinary Perspectives*. New York: Routledge.

Hopkins, Rob. 2014. *The Transition Handbook: From Oil Dependency to Local Resilience*. UIT Cambridge Limited.

Isenhour, Cindy, Mari Martiskainen and Lucie Middlemiss. 2019. *Power and Politics in Sustainable Consumption Research and Practice*. Philadelphia: Routledge. [VitalSource Bookshelf]. Retrieved from https://bookshelf.vitalsource.com/#/books/9781351677301/.

Johnston, Josee. 2008. "The Citizen–Consumer Hybrid: Ideological Tensions and the Case of Whole Foods Market." *Theory and Society*, 37(3): 229–270.

King, Christine A. 2008. "Community Resilience and Contemporary Agri-ecological Systems: Reconnecting People and Food, and People with People." *Systems Research and Behavioral Science*, 25: 111–124.

Kollmuss, A. and J. Agyeman. 2002. "Mind the Gap: Why Do People Act Environmentally and What are the Barriers to Pro-Environmental Behavior?" *Environmental Education Research*, 8(3): 239–260.

Lee, Caroline. 2015. *Democratizing Inequalities: Dilemmas of the New Public Participation*. New York: NYU Press.

Lorek, S. and D. Fuchs. 2013. "Strong Sustainable Consumption Governance–Precondition for a Degrowth Path?" *Journal of Cleaner Production*, 38: 36–43.

Marx, Karl and Friedrich Engels. 1967 [1848]. *The Communist Manifesto*. New York: Monthly Review.

Marx, Karl. 1990 [1867]. *Capital: Volume I*. Translated by Ben Fowkes. London: Penguin Books.

McAdam, Doug. 2000. "Culture and Social Movements." In L. Crothers and C. Lockhart (eds.) *Culture and Politics*. New York: Palgrave Macmillan.

McCarthy, John D. 2013. "Social Movement Sector." In *The Wiley-Blackwell Encyclopedia of Social and Political Movements*. https://doi.org/10.1002/9780470674871.wbespm196.

McCarthy, John D. and Mayer Zald. 1977. "Resource Mobilization and Social Movements: A Partial Theory." *American Journal of Sociology*, 82(6): 1212–1241.

McClintock, Nathan. 2014. "Radical, Reformist, and Garden-Variety Neoliberal: Coming to Terms with Urban Agriculture's Contradictions." *Local Environment*, 19: 147–171.

McPherson, Miller, Lynn Smith-Lovin and James M. Cook. 2001. "Birds of a Feather: Homophily in Social Networks." *Annual Review of Sociology*, 27: 415–444.

Middlemiss, Lucie. 2018. *Sustainable Consumption*. Philadelphia: Routledge. [VitalSource Bookshelf]. Retrieved from https://bookshelf.vitalsource.com/#/books/9781317239819/.

Mincyte, D. and K. Dobernig. 2016. "Urban Farming in the North American Metropolis: Rethinking Work and Distance in Alternative Agro-Food Networks." *Environment and Planning A*, 48(9): 1767–1786.

ORourke, D. and N. Lollo. 2015. "Transforming Consumption: From Decoupling, to Behavior Change, to System Changes for Sustainable Consumption. *Annual Review of Environment and Resources*, 40: 233–259.

Polanyi, Karl. 1944. *The Great Transformation: The Political and Economic Origins of Our Time*. Boston, MA: Beacon Press.

Szasz, Andrew. 2007. *Shopping our Way to Safety: How we Changed from Protecting the Environment to Protecting Ourselves*. Minneapolis: University of Minnesota Press.

Urry, John. 2010. "Consuming the Planet to Excess." *Theory, Culture and Society*, 27(2–30): 191–212.

Weber, Max. 1930. *The Protestant Ethic and the Spirit of Capitalism.* Translated by Talcott Parsons. New York: Scribner.

Whiteman, Gail and William H. Cooper. 2000. "Ecological Embeddedness." *Academy of Management Journal,* 43(6): 1265–1282.

Whitmarsh, L., G. Seyfang and S. O'Neill. 2011. "Public Engagement with Carbon and Climate Change: To What Extent is the Public 'Carbon Capable'?" *Global Environmental Change,* 21(1): 56–65.

Index

Page numbers in bold refer to tables.

Advocates for Urban Agriculture 1, 91, 95
agricultural processes 72
agroecology 69
alienation 41–42, 44, 45, 55–58, 60, 68, 113
Allard, Scott 24
alternative banking schemes 39
alternative currencies 3, 6, 41
alternative economic relationships 106
alternative food: movement 8; networks 41, 48, 69; systems 53
American Ghetto 21
animal-based food production 91
animal product 2
anthropology 7
anti-capitalism 61; *see also* capitalism/capitalist
Auden, W.H. 118

bartering 3, 6, 30, 56, 58, 78, 88, 96, 103–104, 115
Baumann, Shyon 54
Beck, Ulrich 2, 38
Berman, Morris 6, 38, 41
Berman's Dual Process Theory 4, 38, 49n1
Bhutan 41
biodiversity 69
Bowen, S. 54
bureaucracy 42

business-as-usual capitalism 5

capitalism/capitalist 2, 39, 55, 112, 117; economic system 43; exploitation 41–42; market economy 5; and shadow structures 39; society 44; system 39–40
Chicago 20; Advocates for Urban Agriculture 1, 91, 95; Chicken Coop Tour 93, 98–99; Chicken Enthusiasts 93, 97; chicken keepers 27; corrupt politics 21; de-industrialization in 20; egg production 27; field site, geography in 22–24; geographic diversity 23; home food gardens 26; industrial production 25; machine politics 21; methods 30–32; neoliberal late capitalism 20, 22; rural communities 22; rural field site 28–30; SFP 22–23; urban and suburban field sites 24–28; urban communities 22; vegetable gardens 22
chickens 73, 91–92; coop construction 93; coop tour 95; and goats 89
class theory 52
climate change 76
commodification of nature 55, 61–62, 112
communities: collaboration 5; energy

production 39; gardeners 8;
  gardens 86; of practice 92
conservation-minded land
  management 69
conservation of land 69, 71
consumer capitalist culture 6
consumption: of consumer goods 74;
  pattern 52–53
Cooper, William H. 70, 80
co-operative housing 39
countering alienation 41–45,
  52–53, 54
craftsmanship 54
criminology 21
cross-national surveys 32n1
cultural capital 53; consumers 54, 62;
  environmentalist consumers 52;
  phenomenon 54
cultural meanings 11
culture, analyses of 42
Cvetičanin, Predrag 5

Daley, Richard 21
decentralized currency 39
de-commodification 3
degrees of social freedom 7
Democratizing Inequalities (Lee) 100
disenchantment 42, 53, 57, 61
disenchantment of industrialism 55
Dobernig, K. 53
downshifters 3
dual process and shadow
  structures 37–41
Dual Process theory 4, 38, 49n1
dumpster diving 3

ecological caretaking 80
ecological degradation 10
ecological embeddedness 68–70,
  72–73, 80, 114–115; of
  environmentalists 70–75; of non-
  environmentalists 75–80; theory 69
ecological estrangement 69
ecological modernization 5
ecological reciprocity 80
ecological respect 80
economic affiliations 101
economic austerity 40
economic blight 10
economic crisis 6

economic exchanges 96
economic indicators 41
efficacy 89
Emmanuel, Rahm 21
energy decline 57
energy problems 58
entrepreneurial freedoms 46
environmental identities 70
environmentalism 11, 80–82
environmental justice 26
environmental practices 11
environmental problems 57–58
environmental re-interpretations 37
environmental sociology 37
environmental sustainability 5
equality of exchange 5
ethical consumer 54
ethical consumption 52, 54
exploitation of working class 112

Ferrell, Jeff 6
feudalism 39
field sites 23; rural 24; suburban
  24–28; urban 24–28
fisherman 76
fishing 23, 58, 75–76, 81, 89
food: access 10; consumption in high
  season 31; gardening 26; for
  household consumption 46;
  production 2, 74; self-provisioning
  8–9; sustainably 8; tastes 53
"foodie" culture 54, 62
Foster, John Bellamy 43–45
freecycling 3
free market 46
fruit trees 75, 89
fundamental transformation 38

Galt, Ryan E. 37, 45–46, 53
geography 7
Gibson-Graham, J.K. 106
gift economies 3, 39, 41
gleaning public fruit 23
global capitalist system 3
global society 2
Gray, Leslie C. 37
Great Depression 26, 117
Great Recession of 2007/2008 20,
  27, 40
Great Transformation 38, 117

greenhouse gas emitters 68
Growth National Happiness 41
guiding theories 37; countering
    alienation 41–45; dual process and
    shadow structures 37–41; shadow
    structure development, paradox
    in 45–49

harvesting plants 26
Hayes-Conroy, Allison 48, 62
Hayes-Conroy, Jessica 48, 61, 114
health issues 10
healthy ecosystems 80
hegemonic capitalist system 58
hegemonic industrial system 3,
    38–39
herbicide 71
high cultural capital 53
high cultural capital phenomenon
    54
Hochschild, Arlie Russell 81
Holleman, Hannah 45
home-canned food 67
homophilous group 52
Hopkins, Rob 6
household food production 9
Hügelkultur 26
humane hunting practices 80
human–environment interaction 81
human labor 112
hunting 23, 58, 75, 81
Hurley, Patrick 37
hyper-local environment 68

inclusive environmentalism 80–82
income inequality 20
industrial agricultural practices 70
industrial agriculture 43–44, 69
industrial agro-food system 79
industrial capitalism 32, 45, 55
industrial civilization 112
industrialization 2, 58
industrial processes 57
industrial revolution 2, 38
inequality, formation of 42
inverted quarantine 90, 115
Ithaca Hours 6

Johnston, Josee 54

Katz-Gerro, Tally 5–6
Kennedy, Emily H. 54
Kirwan, James 69–71
Klein, Naomi 40
Kotlowitz, Alex 20–21

Lake Michigan 23
landfill diversion 75
large-scale industry 43–44
large vegetable gardens 23
late capitalism 3–4, 21, 39–41, 48,
    106, 113
Lee, Carolyn 99
Leguina, Adrian 5
Lietaer, Bernard 6
livestock 23; *see also* chickens
low cultural capital 53
low-income communities 9, 81

Maguire, Jennifer Smith 62
marginalized environmentalism 4–10,
    67–68, 114–115; ecological
    embeddedness 68–70, 114–115; of
    environmentalists 70–75; of non-
    environmentalists 75–80; inclusive
    environmentalism 80–82
marine environment 77
market forces 2–3
Marx, Karl 2–4, 37–38, 41, 43–45, 53,
    60–61, 68, 112
Marxism: environmental theory
    43–44; theory 10
material reality 112
MaxQDA Qualitative Software 32
McClintock, Nathan 37, 45, 48,
    53, 61–62
McMichael, Phillip 44
meat 2
metabolic rift theory 43
metamorphosis 38
metropolitan area 24
micro-environments 78
Mincyte, D. 27, 53
modern capitalism 38
modernity, transition to 53
modernization 58
modern liberalism 47
Morris, Carol 69–71
multi-pronged mobilization 97

municipal code 27
mushrooms and berries 89

neoliberal capitalism 6; *see also*
  capitalism/capitalist
neoliberalism 21, 48, 61
neoliberal late capitalism 20
neo-Polanyian shadow structure 68
neo-Polanyian theories 38, 55, 62
NGOs 12
non-environmentalists 78
non-liberal values 46
non-metropolitan area 24
non-monetary exchanges/bartering 3

organic pest/disease management
  69, 75

Paxon, Heather 8
permaculture 26
Polanyi, Karl 2, 4, 37–38, 53, 112
Polanyian double movement 39, 86,
  90, 101, 105
political environmentalism 67
political mobilization 96
politics of perfection 62
poultry, banning 96
practitioner networks 86, 92,
  115–117; SFP 86–105; shadow
  structures 105–107
pre-modern society 101–103
privatization of public resources 21
production, act of 52–53, 55, 90
pro-environmental behaviors 74
pro-environmental innovation 81
purposive sampling 30–32

radical countermovement 112
rationalization 42
recipes, swapping 26
recycling 74
renewable technologies 5
Rieger, Jeorg 39–40
rotating crops 74

safety 89
Sahakian, Marlyne 5
sampling: descriptive statistics **31**; to
  maximize range 30
Schoolman, Ethan D. 54

Schor, Juliet B. 5, 7
self-help 46
self-identified environmentalists 12
self-production of food 6, 11, 55, 58,
  72, 76, 82, 90, 95, 105
self-regulating market 3
self-reliance 89
self-sufficiency 46
self-sustaining 89
seminal theory of double
  movement 2–3
sense of meaninglessness 58
SFP *see* subsistence food
  production (SFP)
shadow structures 3, 37, 39, 49n1,
  115–117; and capitalism 39;
  dual process and 37–41;
  economic 39; paradox in
  development 45–49; process of 41;
  in Puerto Rico 40
sharing 6
sharing economies 5, 7, 39
Shopping Our Way to Safety
  (Szasz) 89–90
small-scale agriculture 9
small-scale food production 7, 26, 39
small-scale sustainable agriculture 9
social cohesion 5
social difference 52
social empowerment 47
socioeconomic system 38
sociology 41
state-authoritarian structure 48
Strangers in their Own Land
  (Hochschild) 81
subsistence food producers (SFPers)
  23, 30–32, 52, 68–69, 106, 113
subsistence food production (SFP) 2,
  7, 12n1, 37, 45, 48; animal
  agriculture 2; Chicago 22, 113;
  dimensions of difference and
  developing ties 91–96; motivation
  for 56–58; populations 9, 53;
  practices 71; shadow structure
  development 96–105; social change
  86–88; social problems 88–91
subsistence production 44
suburban field sites 24–28
suburban municipality 27
suburbs 24

sustainable consumption 3–6, 9, 52, 74, 89
sustainable land use 71
sustainable use of land 69
Szasz, Andrew 89–90

theory of Interstitial Transformation 47
There Are No Children Here (Kotlowitz) 20
time banks 6, 41
tool or seed libraries 3
Transition Towns 6, 41
transportation, alternative modes of 3

United Nations 9
urban agriculture 8, 10
Urban Agriculture Livestock Expo 1–2, 91, 93
urban field sites 24–28
urban planning 7
Urry, John 38, 117

USDA ERS Food Security Survey Module 32n3

values-system 71
vegetable gardening 75, 89
viscerality 52, 54, 58–61, 112
voluntary simplicity 3

waste processing 75
watering 74
Weber, Max 2–3, 37, 41–42, 45, 53, 61, 101, 112; concept of disenchantment 68, 73; modern and pre-modern Western society **42**, 42–43, 103; *see also* industrial civilization
Weberian environmentalism 42
Wengronowitz, Robert 5, 7
Whiteman, Gail 70, 80
white working-class suburb 91
work for self-producing 76
working-class consumer 54
World Bank 9
Wright, Erik Olin 37, 45, 47–48

Printed in the United States
by Baker & Taylor Publisher Services